Welded design – theory and practice

John Hicks

William Andrew
publishing

ABINGTON PUBLISHING
Cambridge England

Published by Abington Publishing
Woodhead Publishing Limited, Abington Hall,
Abington, Cambridge CB1 6AH, England
www.woodhead-publishing.com

Published in North and South America by William Andrew Inc.,
13 Eaton Avenue, Norwich, NY 13815, USA

First published 2001, Abington Publishing

British Library Cataloguing in Publication Data
A catalogue record for this book is available from the British Library.

ISBN 1 85573 537 7 (Woodhead Publishing Limited)
ISBN 0-8155-1474-3 (William Andrew Inc.)

Cover design by The ColourStudio
Typeset by BookEns Ltd, Royston, Herts
Printed by T J International, Cornwall, England

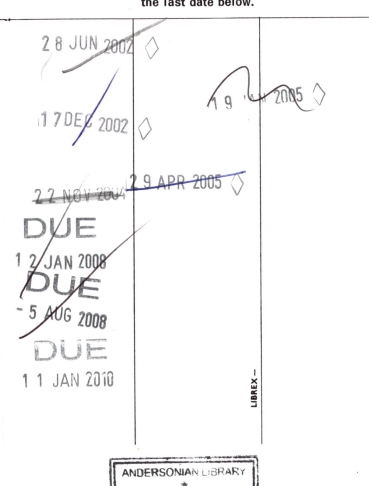

Welded design –
theory and practice

Contents

I have written this book for engineers of all disciplines, and this includes those welding engineers who do not have a background in matters of engineering design, as well as for others in all professions who may find this subject of interest. As might be expected, I have drawn heavily on my own experience. Not that I discovered any new principles or methods but because I had the privilege of firstly being associated with research into the behaviour of welded joints in service at its most active time in the 1960s and 1970s and secondly with the application of that research in a range of industries and particularly in structural design and fabrication which accompanied the extension of oil and gas production into deeper waters in the 1970s. The results of those developments rapidly spread into other fields of structural engineering and I hope that this book will be seen in part as a record of some of the intense activity which went on in that period, whether it was in analysing test results in a laboratory, writing standards, preparing a conceptual design or installing a many thousand tonne substructure on the ocean floor.

The position from which I write this book is one where, after being a structural engineer for five years, I became a specialist in welded design. In this role I have for many years worked with colleagues, clients and pupils who, without exception, have been and are a pleasure to work with; their mastery of their own disciplines and the responsibilities which they carry dwarfs my own efforts. I have also spent, I believe, sufficient periods in other occupations both inside and outside the engineering profession to give me an external perspective on my specialism. As a result I felt that it would be helpful to write a book setting out the subject of welded design in the context of the overall picture of engineering with some historical background. In presenting the subject in this way I hope that it will encourage teaching staff in universities and colleges to see welded joints and their behaviour as an integral part of engineering and that they will embed the subject in their courses instead of treating it as an add-on. It will also serve practising welding and other engineers wishing to extend their knowledge of the opportunities which welding offers and the constraints it imposes in their own work.

The subject of design for welding rests at a number of interfaces between the major engineering disciplines as well as the scientific disciplines of physics, chemistry and metallurgy. This position on the boundaries between traditional mainstream subjects may perhaps be the reason why it receives relatively little attention in university engineering courses at undergraduate level. My recent discussions with engineering institutions and academics reveals a situation, both in the UK and other countries, in which the appearance or otherwise of the subject in a curriculum seems to depend on whether or not there is a member of the teaching staff who has both a particular interest in the subject and can find the time in the timetable. This is not a new position; I have been teaching in specialist courses on design for welding at all academic and vocational levels since 1965 and little seems to have changed. Mr R P Newman, formerly Director of Education at The Welding Institute, writing in 1971,[1] quoted a reply to a questionnaire sent to industry:

> *Personnel entering a drawing office without much experience of welding, as many do today* (i.e. 1971)*, can reach a reasonably senior position and still have only a 'stop-gap' knowledge, picked up on a general basis. This is fundamentally wrong and is the cause of many of our fabrication/design problems.*

There was then, and has been in the intervening years, no shortage of books and training courses on the subject of welded design but the matter never seems to enter or remain in many people's minds. In saying this I am not criticising the individual engineers who may have been led to believe that welded joint design and material selection are matters which are either not part of the designer's role or, if they are, they require no education in the subjects. Indeed, such was my own early experience in a design office and I look back with embarrassment at my first calculation of the suitability of welded joint design in an industry in which welding was not commonly used. It was an example of being so ignorant that I didn't know that I was ignorant. That first experience of a premature failure has stayed with me and gives me humility when assisting people who are in a similar position today. 'There, but for the grace of God, go I' should be on a banner above every specialist's desk. There are, of course many engineers who have, either because their work required it or because of a special interest, become competent in the subject. Either way, there is a point at which a specialist input is required which will depend upon the nature, novelty and complexity of the job set against the knowledge and experience of the engineer.

I have tried to put into this book as much as is useful and informative without including a vast amount of justification and detail; that can be found in the referenced more specialist works. However, I have tried to keep a balance in this because if too many matters are the subject of references the reader may become exasperated at continually having to seek other books, some of which will be found only in specialist libraries. For the most part I have avoided references to standards and codes of practice except in a historical context. Exceptions are where a standard is an example of basic design data or where it represents guidance on an industry wide agreed approach to an analytical process. I have adopted this position because across the world there are so many standards and they are continually being amended. In addition standards do not represent a source of fundamental knowledge although, unfortunately, some are often seen in that light. However I recognise their importance to the practical business of engineering and I devote a chapter to them.

I acknowledge with pleasure those who have kindly provided me with specialist comment on some parts of the book, namely Dr David Widgery of ESAB Group (UK) Ltd on welding processes and Mr Paul Bentley on metallurgy. Nonetheless I take full responsibility for what is written here. I am indebted to Mr Donald Dixon CBE for the illustration of the Cleveland Colossus North Sea platform concept which was designed when he was Managing Director of The Cleveland Bridge and Engineering Co Ltd. For the photographs of historic structures I am grateful to the Chambre de Commerce et d'Industrie de Nîmes, the Ironbridge Gorge Museum, and Purcell Miller Tritton and Partners. I also am pleased to acknowledge the assistance of TWI, in particular Mr Roy Smith, in giving me access to their immense photographic collection.

JOHN HICKS

Many engineering students and practising engineers find materials and metallurgy complicated subjects which, perhaps amongst others, are rapidly forgotten when examinations are finished. This puts them at a disadvantage when they need to know something of the behaviour of materials for further professional qualifications or even their everyday work. The result of this position is that engineering decisions at the design stage which ought to take account of the properties of a material can be wrong, leading to failures and even catastrophes. This is clearly illustrated in an extract from *The Daily Telegraph* on 4 September 1999 in an article offering background to the possible cause of a fatal aircraft crash. ' "There is no fault in the design of the aircraft," the (manufacturer's) spokesman insisted. "It is a feature of the material which has shown it does not take the wear over a number of years. . ." ' This dismissal of the designer's responsibility for the performance of materials is very different in the case of concrete in which every civil engineer appears to have been schooled in its constituent raw materials, their source, storage, mixing, transport and pouring as well as the strength.

To emphasise the wider responsibility which the engineer has I give the background to some of the materials and the techniques which the engineer uses today and make the point that many of the design methods and data in common use are based on approximations and have limitations to their validity. A number of so-called rules have been derived on an empirical basis; they are valid only within certain limits. They are not true laws such as those of Newtonian mechanics which could be applied in all terrestrial and some universal circumstances and whose validity extends even beyond the vision of their author himself; albeit Newton's laws have been modified, if not superseded, by Einstein's even more fundamental laws.

The title of this book reflects this position for it has to be recognised that there is precious little theory in welded joint design but a lot of practice. There appear in this book formulae for the strength of fillet welds which look very theoretical whereas in fact they are empirically derived from large numbers of tests. Similarly there are graphs of fatigue life which look mathematically based but are statistically derived lines of the probability of failure of test specimens from hundreds of fatigue tests; subsequent theoretical work in the field of fracture mechanics has explained why the graphs have the slope which they do but we are a long way from being able to predict on sound scientific or mathematical grounds the fatigue life of a particular item as a commonplace design activity. Carbon equivalent formulae are attempts to quantify the weldability of steels in respect of hardenability of the heat affected zone and are examples of the empirical or arbitrary rules or formulae surrounding much of welding design and fabrication. Another example, not restricted to welding by any means, is in fracture mechanics which uses, albeit in a mathematical context, the physically meaningless unit $Nmm^{-3/2}$. Perhaps in the absence of anything better we should regard these devices as no worse than a necessary and respectable mathematical fudge – perhaps an analogy of the cosmologist's black hole.

A little history helps us to put things in perspective and often helps us to understand concepts which otherwise are difficult to grasp. The historical background to particular matters is important to the understanding of the engineer's contribution to society, the way in which developments take place and the reasons why failures occur. I have used the history of Britain as a background but this does not imply any belief on my part that history elsewhere has not been relevant. On one hand it is a practical matter because I am not writing a history book and my references to history are for perspective only and it is convenient to use that which I know best. On the other hand there is a certain rationale in using British history in that Britain was the country in which the modern industrial revolution began, eventually spreading through the European continent and elsewhere and we see that arc welding processes were the subject of development in a number of countries in the late nineteenth century. The last decade of the twentieth century saw the industrial base move away from the UK, and from other European countries, mainly to countries with lower wages. Many products designed in European countries and North America are now manufactured in Asia. However in some industries the opposite has happened when, for example, cars designed in Japan have been manufactured for some years in the UK and the USA. A more general movement has been to make use of manufacturing capacity and specialist processes wherever they are available. Components for some US aircraft are made in Australia, the UK and other countries; major components for some UK aircraft are made in Korea. These are only a few examples of a general trend in which manufacturing as well as trade is becoming global. This dispersion of industrial activity makes it important that an adequate understanding of the relevant technology exists across the globe and this must include welding and its associated activities.

Not all engineering projects have been successful if measured by conventional commercial objectives but some of those which have not met these objectives are superb achievements in a technical sense. The Concorde airliner and the Channel Tunnel are two which spring to mind. The Concorde is in service only because its early development costs were underwritten by the UK and French governments. The Channel Tunnel linking England and France by rail has had to be re-financed and its payback time rescheduled far beyond customary periods for returns on investment. Further, how do we rate the space programmes? Their payback time may run into decades, if not centuries, if at all. Ostensibly with a scientific purpose, the success of many space projects is more often measured not in scientific or even commercial terms but in their political effect. The scientific results could often have been acquired by less extravagant means. In defence equipment, effectiveness and reliability under combat conditions, possibly after lengthy periods in storage, are the prime requirements here although cost must also be taken into account.

There are many projects which have failed to achieve operational success through lack of commitment, poor performance, or through political interference. In general their human consequences have not been lasting. More sadly there are those failures which have caused death and injury. Most of such engineering catastrophes have their origins in the use of irrelevant or invalid methods of analysis, incomplete information or the lack of understanding of material behaviour, and, so often, lack of communication. Such catastrophes are relatively rare, although a tragedy for those involved.

What is written in this book shows that accumulated knowledge, derived over the years from research and practical experience in welded structures, has been incorporated into general design practice. Readers will not necessarily find herein all the answers but I hope that it will cause them to ask the right questions. The activity of engineering design calls on the knowledge of a variety of engineering disciplines many of which have a strong theoretical, scientific and intellectual background leavened with some rather arbitrary adjustments and assumptions. Bringing this knowledge to a useful purpose by using materials in an effective and economic way is one of the skills of the engineer which include making decisions on the need for and the positioning of joints, be they permanent or temporary, between similar or dissimilar materials which is the main theme of this book. However as in all walks of engineering the welding designer must be aware that having learned his stuff he cannot just lean back and produce designs based on that knowledge. The world has a habit of changing around us which leads not only to the need for us to recognise the need to face up to demands for new technology but also being aware that some of the old problems revisit us. Winston Churchill is quoted as having said that the further back you look the further forward you can see.

1
The engineer

1.1 Responsibility of the engineer

As we enter the third millennium *annis domini*, most of the world's population continues increasingly to rely on man-made and centralised systems for producing and distributing food and medicines and for converting energy into usable forms. Much of these systems relies on the, often unrecognised, work of engineers. The engineer's responsibility to society requires that not only does he keep up to date with the ever faster changing knowledge and practices but that he recognises the boundaries of his own knowledge. The engineer devises and makes structures and devices to perform duties or achieve results. In so doing he employs his knowledge of the natural world and the way in which it works as revealed by scientists, and he uses techniques of prediction and simulation developed by mathematicians. He has to know which materials are available to meet the requirements, their physical and chemical characteristics and how they can be fashioned to produce an artefact and what treatment they must be given to enable them to survive the environment.

The motivation and methods of working of the engineer are very different from those of a scientist or mathematician. A scientist makes observations of the natural world, offers hypotheses as to how it works and conducts experiments to test the validity of his hypothesis; thence he tries to derive an explanation of the composition, structure or mode of operation of the object or the mechanism. A mathematician starts from the opposite position and evolves theoretical concepts by means of which he may try to explain the behaviour of the natural world, or the universe whatever that may be held to be. Scientists and the mathematicians both aim to seek the truth without compromise and although they may publish results and conclusions as evidence of their findings their work can never be finished. In contrast the engineer has to achieve a result within a specified time and cost and rarely has the resources or the time to be able to identify and verify every possible piece of information about the environment in which the artefact has to operate or the response of the artefact to that environment. He has to work within a degree of uncertainty, expressed by the probability that the artefact will do what is expected of it at a defined cost and for a specified life. The engineer's circumstance is perhaps summarised best by the oft quoted request: 'I don't want it perfect, I want it Thursday!' Once the engineer's work is complete he cannot go back and change it without disproportionate consequences; it is there for all to see and use. The ancient Romans were particularly demanding of their bridge engineers; the engineer's name had to be carved on a stone in the bridge, not to praise the engineer but to know who to execute if the bridge should collapse in use!

People place their lives in the hands of engineers every day when they travel, an activity associated with which is a predictable probability of being killed or injured by the omissions of their fellow drivers, the mistakes of professional drivers and captains or the failings of the engineers who designed, manufactured and maintained the mode of transport. The engineer's role is to be seen not only in the vehicle itself, whether that be on land, sea or air, but also in the road, bridge, harbour or airport, and in the navigational aids which abound and now permit a person to know their position to within a few metres over and above a large part of the earth.

Human error is frequently quoted as the reason for a catastrophe and usually means an error on the part of a driver, a mariner or a pilot. Other causes are often lumped under the catch-all category of *mechanical failure* as if such events were beyond the hand of man; a naïve attribution, if ever there were one, for somewhere down the line people were involved in the conception, design, manufacture and maintenance of the device. It is therefore still human error which caused the problem even if not of those immediately involved. If we need to label the cause of the catastrophe, what we should really do is to place it in one of, say, four categories, all under the heading of human error, which would be failure in specification, design, operation or maintenance. An 'Act of God' so beloved by judges is a get-out. It usually means a circumstance or set of circumstances which a designer, operator or legislator ought to have been able to predict and allow for but chose to ignore. If this seems very harsh we have only to look at the number of lives lost in bulk carriers at sea in the past years. There still seems to be a culture in seafaring which accepts that there are unavoidable hazards and which are reflected in the nineteenth century hymn line '. . . for those in peril on the sea'. Even today there are cultures in some countries which do not see death or injury by man-made circumstances as preventable or even needing prevention; concepts of risk just do not exist in some places. That is not to say that any activity can be free of hazards; we are exposed to hazards throughout our life. What the engineer should be doing is to conduct activities in such a way that the probability of not surviving that hazard is known and set at an accepted level for the general public, leaving those who wish to indulge in high risk activities to do so on their own.

We place our lives in the hands of engineers in many more ways than these obvious ones. When we use domestic machines such as microwave ovens with their potentially injurious radiation, dishwashers and washing machines with a potentially lethal 240 V supplied to a machine running in water into which the operator can safely put his or her hands. Patients place their lives in the hands of engineers when they submit themselves to surgery requiring the substitution of their bodily functions by machines which temporarily take the place of their hearts, lungs and kidneys. Others survive on permanent replacements for their own bodily parts with man-made implants be they valves, joints or other objects. An eminent heart surgeon said on television recently that heart transplants were simple; although this was perhaps a throwaway remark one has to observe that if it is simple for him, which seems unlikely, it is only so because of developments in immunology, on post-operative critical care and on anaesthesia (not just the old fashioned gas but the whole substitution and maintenance of complete circulatory and pulmonary functions) which enables it to be so and which relies on complex machinery requiring a high level of engineering skill in design, manufacturing and maintenance. We place our livelihoods in the hands of engineers who make machinery whether it be for the factory or the office.

Businesses and individuals rely on telecommunications to communicate with others and for some it would seem that life without television and a mobile telephone would be at best meaningless and at worst intolerable. We rely on an available supply of energy to enable us to use all of this equipment, to keep ourselves warm and to cook our food. It is the engineer who converts the energy contained in and around the Earth and the Sun to produce this supply of usable energy to a remarkable level of reliability and consistency be it in the form of fossil fuels or electricity derived from them or nuclear reactions.

1.2   Achievements of the engineer

The achievements of the engineer during the second half of the twentieth century are perhaps most popularly recognised in the development of digital computers and other electronically based equipment through the exploitation of the discovery of semi-conductors, or transistors as they came to be known. The subsequent growth in the diversity of the use of computers could hardly have been expected to have taken place had we continued to rely on the thermionic valve invented by Sir Alexander Fleming in 1904, let alone the nineteenth century mechanical calculating engine of William Babbage. However let us not forget that at the beginning of the twenty-first century the visual displays of most computers and telecommunications equipment still rely on the technology of thermionic emission. The liquid crystal has occupied a small area of application and the light emitting diode has yet to reach its full potential.

The impact of electronic processing has been felt both in domestic and in business life across the world so that almost everybody can see the effect at first hand. Historically most other engineering achievements probably have had a less immediate and less personal impact than the semi-conductor but have been equally significant to the way in which trade and life in general was conducted. As far as life in the British Isles was concerned this process of accelerating change made possible by the engineer might perhaps have begun with the building of the road system, centrally heated villas and the setting up of industries by the Romans in the first few years AD. However their withdrawal 400 years later was accompanied by the collapse of civilisation in Britain. The invading Angles and Saxons enslaved or drove the indigenous population into the north and west; they plundered the former Roman towns and let them fall into ruin, preferring to live in small self-contained settlements. In other countries the Romans left a greater variety of features; not only roads and villas but mighty structures such as that magnificent aqueduct, the Pont du Gard in the south of France (Fig. 1.1). Hundreds of years were to pass before new types of structures were erected and of these perhaps the greatest were the cathedrals built by the Normans in the north of France and in England. The main structure of these comprised stone arches supported by external buttresses in between

1.1   The Pont du Gard (photograph by Bernard Liegeois).

which were placed timber beams supporting the roof. Except for these beams all the material was in compression. The modern concept of a structure with separate members in tension, compression and shear which we now call chords, braces, ties, webs, etc. appears in examples such as Ely Cathedral in the east of England. The cathedral's central tower, built in the fourteenth century, is of an octagonal planform supported on only eight arches. This tower itself supports a timber framed structure called the lantern (Fig. 1.2). However let us not believe that the engineers of those days were always successful; this octagonal tower and lantern at Ely had been built to replace the Norman tower which collapsed in about 1322.

Except perhaps for the draining of the Fens, also in the east of England, which was commenced by the Dutch engineer, Cornelius Vermuyden, under King Charles I in 1630, nothing further in the modern sense of a regional or national infrastructure was developed in Britain until the building of canals in the eighteenth century. These were used for moving bulk materials needed to feed the burgeoning industrial revolution and the motive power was provided by the horse. Canals were followed by, and to a great extent superseded by, the railways of the nineteenth century powered by steam which served to carry both goods and passengers, eventually in numbers, speed and comfort which the roads could not offer. Alongside these came the emergence of the large oceangoing ship, also driven by steam, to serve the international trade in goods of all types. The contribution of the inventors and developers of the steam engine, initially used to pump water from mines, was therefore central to the growth of transport. Amongst them we acknowledge Savory, Newcomen, Trevithick, Watt and Stephenson. Alongside these developments necessarily grew the industries to build the means and to make the equipment for transport and which in turn provided a major reason for the existence of a transport system, namely the production of goods for domestic and, increasingly, overseas consumption.

Today steam is still a major means of transferring energy in both fossil fired and nuclear power stations as well as in large ships using turbines. Its earlier role in smaller stationary plant and in other transport applications was taken over by the internal combustion engine both in its piston and turbine forms. Subsequently the role of the stationary engine has been taken over almost entirely by the electric motor. In the second half of the twentieth century the freight carrying role of the railways became substantially subsumed by road vehicles resulting from the building of motorways and increasing the capacity of existing main roads (regardless of the wider issues of true cost and environmental damage). On a worldwide basis the development and construction of even larger ships for the cheap long distance carriage of bulk materials and of larger aircraft for providing cheap travel for the masses were two other achievements. Their use built up comparatively slowly in the second half of the century but their actual

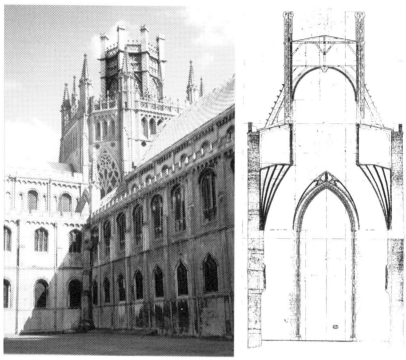

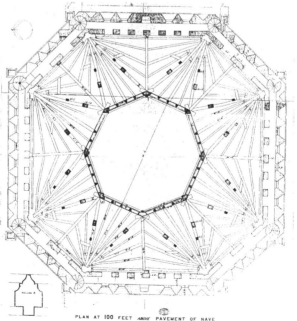

PLAN AT 100 FEET *ABOVE* PAVEMENT OF NAVE

1.2
The lantern of Ely Cathedral (photograph by Janet Hicks, drawings by
courtesy of Purcell Miller Tritton and Partners).

development had taken place not in small increments but in large steps. The motivation for the ship and aircraft changes was different in each case. A major incentive for building larger ships was the closure of the Suez Canal in 1956 so that oil tankers from the Middle East oil fields had to travel around the Cape of Good Hope to reach Europe. The restraint of the canal on vessel size then no longer applied and the economy of scale afforded by large tankers and bulk carriers compensated for the extra distance. The development of a larger civil aircraft was a bold commercial decision by the Boeing Company. Its introduction of the type 747 in the early 1970s immediately increased the passenger load from a maximum of around 150 to something approaching 400. In another direction of development at around the same time British Aerospace (or rather, its predecessors) and Aérospatiale offered airline passengers the first, and so far the only, means of supersonic travel. Alongside these developments were the changes in energy conversion both to nuclear power as well as to larger and more efficient fossil-fuelled power generators. In the last third of the century extraction of oil and gas from deeper oceans led to very rapid advancements in structural steel design and in materials and joining technologies in the 1970s. These advances have spun off into wider fields of structural engineering in which philosophies of structural design addressed more and more in a formal way matters of integrity and economy. In steelwork design generally more rational approaches to probabilities of occurrences of loads and the variability of material properties were considered and introduced. These required a closer attention to questions of quality in the sense of consistency of the product and freedom from features which might render the product unable to perform its function.

1.3   The role of welding

Bearing in mind the overall subject of this book we ought to consider if and how welding influenced these developments. To do this we could postulate a 'what if?' scenario: what if welding had not been invented? This is not an entirely satisfactory approach since history shows that the means often influences the end and vice versa; industry often maintains and improves methods which might be called old fashioned. As an example, machining of metals was, many years ago, referred to by a proponent of chemical etching as an archaic process in which one knocks bits off one piece of metal with another piece of metal, not much of an advance on Stone Age flint knapping. Perhaps this was, and still is, true; nonetheless machining is still widely used and shaping of metals by chemical means is still a minority process. Rivets were given up half a century ago by almost all industries except the aircraft industry which keeps them because they haven't found a more suitable way of joining their chosen materials; they make a very good job of it, claiming the benefit over welding of a structure with natural crack stoppers. As a confirmation of its integrity a major joint in a Concorde fuselage was taken apart after 20 years' service and found to be completely sound. So looking at the application of welding there are a number of aspects which we could label feasibility, performance and costs. It is hard to envisage the containment vessel of a nuclear reactor or a modern boiler drum or heat exchanger being made by riveting any more than we could conceive of a gas or oil pipeline being made other than by welding. If welding hadn't been there perhaps another method would have been used, or perhaps welding would have been invented for the purpose. It does seem highly likely that the low costs of modern shipbuilding, operation, modification and repair can be attributed to the lower costs of welded fabrication of large plate structures over riveting in addition to which is the weight saving. As early as 1933 the editor of the first edition of *The Welding Industry* wrote '... the hulls of German pocket battleships are being fabricated entirely with welding – a practice which produces a weight saving of 1 000 tons per ship'. The motivation for this attention to weight was that under the Treaty of Versailles after the First World War Germany was not allowed to build warships of over 10 000 tons. A year later, in 1934, a writer in the same journal visited the works of A V Roe in Manchester, forerunner of Avro who later designed and built many aircraft types including the Lancaster, Lincoln, Shackleton and Vulcan. 'I was prepared to see a considerable amount of welding, but the pitch of excellence to which Messrs A V Roe have brought oxy-acetylene welding in the fabrication of fuselages and wings, their many types of aircraft and the number of welders that were being employed simultaneously in this work, gave me, as a welding engineer, great pleasure to witness.' The writer was referring to steel frames which today we might still see as eminently weldable. However the scope for welding in airframes was to be hugely reduced in only a few years by the changeover in the later 1930s from fabric covered steel frames to aluminium alloy monocoque structures comprising frames, skin and stringers for the fuselage and spars, ribs and skin for the wings and tail surfaces. This series of alloys was unsuitable for arc welding but resistance spot welding was used much later for attaching the lower fuselage skins of the Boeing 707 airliner to the frames and stringers as were those of the Handley Page Victor and Herald aircraft. The material used, an Al–Zn–Mg alloy, was amenable to spot welding but controls were placed on hardness to avoid stress corrosion cracking. It cannot be said that without welding these aircraft would not have been made, it was just another suitable joining process. The Bristol T188 experimental supersonic aircraft of the late 1950s had an airframe made of TIG spot-welded austenitic stainless steel. This material was chosen for its ability to maintain its strength at the temperatures developed by aerodynamic friction in supersonic flight, and it also happened to be weldable. It was not a solution which was eventually adopted for the Concorde in which a riveted aluminium alloy structure is used but whose temperature is moderated by cooling it with the engine fuel. Apart from these examples and the welded steel tubular space frames formerly used in light fixed wing aircraft and helicopters, airframes have been riveted and continue to be so. In contrast many aircraft engine components are made by welding but gas turbines always were and so the role of welding in the growth of aeroplane size and speed is not so specific. In road vehicle body and white goods manufacture, the welding developments which have supported high production rates and accuracy of fabrication have been as much in the field of tooling, control and robotics as in the welding processes themselves. In construction work, economies are achieved through the use of shop-welded frames or members which are bolted together on site; the extent of the use of welding on site varies between countries. Mechanical handling and construction equipment have undoubtedly benefited from the application of welding; many of the machines in use today would be very cumbersome, costly to make and difficult to maintain if welded assemblies were not used. Riveted road and rail bridges are amongst items which are a thing of the past having been succeeded by welded fabrications; apart from the weight saving, the simplicity of line and lack of lap joints makes protection from corrosion easier and some may say that the appearance is more pleasing.

An examination of the history of engineering will show that few objects are designed from scratch; most tend to be step developments from the previous item. Motor cars started off being called 'horseless carriages' which is exactly what they were. They were horse drawn carriages with an engine added; the shafts were taken off and steering effected by a tiller. Even now 'dash board' remains in everyday speech revealing its origins in the board which protected the driver from the mud and stones thrown up by the horse's hooves. Much recent software for personal computers replicates the physical features of older machinery in the 'buttons', which displays an extraordinary level of conservatism. A similar conservatism can be seen in the adoption of new joining processes. The first welded ships were just welded versions of the riveted construction. It has taken decades for designers to stop copying castings by putting little gussets on welded items. However it can be observed that once a new manufacturing technique is adopted, and the works practices, planning and costing adjusted to suit, it will tend to be used exclusively even though there may be arguments for using the previous processes in certain circumstances.

1.4   Other materials

Having reflected on these points our thoughts must not be trammelled by ignorance of other joining processes or indeed by materials other than the metals which have been the customary subjects of welding. This book concentrates on arc welding of metals because there must be a limit to its scope and also because that is where the author's experience lies. More and more we see other metals and non-metals being used successfully in both traditional and novel circumstances and the engineer must be aware of all the relevant options.

1.5   The welding engineer as part of the team

As in most other professions there are few circumstances today where one person can take all the credit for a particular achievement although a leader is essential. Most engineering projects require the contributions of a variety of engineering disciplines in a team. One of the members of that team in many products or projects is the welding engineer. The execution of the responsibilities of the welding engineer takes place at the interface of a number of conventional technologies. For contributing to the design of the welded product these include structural and mechanical engineering, material processing, weldability and performance and corrosion science. For the setting up and operation of welding plant they include electrical, mechanical and production engineering, the physics and chemistry of gases. In addition, the welding engineer must be familiar with the general management of industrial processes and personnel as well as the health and safety aspects of the welding operations and materials.

Late twentieth century practice in some areas would seem to require that responsibility for the work be hidden in a fog of contracts, sub-contracts and sub-sub-contracts *ad infinitum* through which are employed conceptual designers, detail designers, shop draughtsmen, quantity surveyors, measurement engineers, approvals engineers, specification writers, contract writers, purchasing agencies, main contractors, fabricators, sub-fabricators and inspection companies. All these are surrounded by underwriters and their warranty surveyors and loss adjusters needed in case of an inadequate job brought about by awarding contracts on the basis of price and not on the ability to do the work. Responsibilities become blurred and it is important that engineers of each discipline are at least aware of, if not familiar with, their colleagues' roles.

2.1 Steels

2.1.1 The origins of steel

The first iron construction which makes use of structural engineering principles was a bridge built by Abraham Darby in 1779 over a gorge known as Coalbrookdale through which runs the River Severn at a place named after it, Ironbridge, in Shropshire in the UK (Fig. 2.1). It was in this area that Darby's grandfather had, in 1709, first succeeded in smelting iron with coke rather than charcoal, a technique which made possible the mass production of iron at an affordable price. The bridge is in the form of frames assembled from cast iron bars held together by wedges, a technique carried over from timber construction. Cast iron continued to be used for bridges into the nineteenth century until Robert Stephenson's bridge over the River Dee at Chester collapsed under a train in 1847 killing five people. Although the tension loads were taken by wrought iron bars the bridge failed at their attachment to the cast iron. At the time of that event Stephenson was constructing the Newcastle High Level bridge using cast iron. However he took great care in designing the bow and string girders resting on five stone piers 45 m above the River Tyne so that excessive tension was avoided. The spans are short, the members massive and particular care was taken over their casting and testing. Work commenced on the bridge in 1846 and was completed in three years; it stands to this day carrying road and rail traffic on its two decks. Nevertheless public outcry at the Dee tragedy caused the demise of cast iron for bridge building; its place was taken by wrought iron, which is almost pure iron and a very ductile material, except for members in compression such as columns.

Steels discovered thousands of years ago acquired wide usage for cutlery, tools and weapons; a heat treatment comprising quenching and tempering was applied as a means of adjusting the hardness, strength and toughness of the steel. Eventually steels became one of the most common group of metals

2.1 Ironbridge (photograph by courtesy of the Ironbridge Gorge Museum).

in everyday use and in many ways they are the most metallurgically complex.

Crude iron, or pig iron as it is known, is usually made by smelting iron ore with coke and limestone. It has a high carbon content which makes it brittle and so it is converted to mild steel by removing some or most of the carbon. This was first done on a large industrial scale using the converter invented by Henry Bessemer who announced his process to the British Association in 1856. Some say that he based his process on a patent of James Naysmith in which steam was blown through the molten iron to remove carbon; others held that he based it on the 'pneumatic method', invented two years earlier by an American, William Kelly. Nevertheless it was the Bessemer process that brought about the first great expansion of the British and American steel industries, largely owing to the mechanical superiority of Bessemer's converter.

Developments in industrial steelmaking in the latter part of the nineteenth century and in the twentieth century lead to the present day position where with fine adjustment of the steel composition and microstructure it is possible to provide a wide range of weldable steels having properties to suit the range of duties and environments called upon. This book does not aim to teach the history and practice of iron and steel making; that represents a fascinating study in its own right and the reader interested in such matters should read works by authors such as Cottrell.[2]

The ability of steel to have its properties changed by heat treatment is a valuable feature but it also makes the joining of it by welding particularly complicated. Before studying the effects of the various welding processes on steel we ought to see, in a simple way, how iron behaves on its own.

2.1.2 The atomic structure of iron

The iron atom, which is given the symbol Fe, has an atomic weight of 56 which compares with aluminium, Al, at 27, lead, Pb, at 207 and carbon, C, at 12. In iron at room temperature the atoms are arranged in a regular pattern, or lattice, which is called *body centred cubic* or bcc for short. The smallest repeatable three dimensional pattern is then a cube with an atom at each corner plus one in the middle of the cube. Iron in this form is called ferrite (Fig. 2.2(a)).

(a)

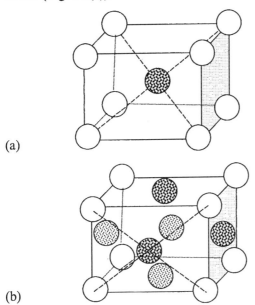

(b)

2.2 (a) Body centred cubic structure; (b) face centred cubic structure.

If iron is heated to 910°C, almost white hot, the layout of the atoms in the lattice changes and they adopt a pattern in which one atom sits in the middle of each planar square of the old bcc pattern. This new pattern is called *face centred cubic*, abbreviated to fcc. Iron in this form is called austenite (Fig. 2.2(b)).

When atoms are packed in one of these regular patterns the structure is described as crystalline. Individual crystals can be seen under a microscope as grains the size of which can have a strong effect on the mechanical properties of the steel. Furthermore some important physical and metallurgical changes can be initiated at the boundaries of the grains. The change from one lattice pattern to another as the temperature changes is called a *transformation*. When iron transforms from ferrite (bcc) to austenite (fcc) the atoms become more closely packed and the volume per atom of iron changes which generates internal stresses during the transformation. Although the fcc pattern is more closely packed the spaces between the atoms are larger than in the bcc pattern which, we shall see later, is important when alloying elements are present.

2.1.3 Alloying elements in steel

The presence of more than about 0·1% by weight of carbon in iron forms the basis of the modern structural steels. Carbon atoms sit between the iron atoms and provide a strengthening effect by resisting relative movements of the rows of atoms which would occur when the material yields. Other alloying elements with larger atoms than carbon can actually take the place of some of the iron atoms and increase the strength above that of the simple carbon steel; the relative strengthening effect of these various elements may differ with temperature. Common alloying elements are manganese, chromium, nickel and molybdenum, which may in any case have been added for other reasons, e.g. manganese to combine with sulphur so preventing embrittlement, chromium to impart resistance to oxidation at high temperatures, nickel to increase hardness, and molybdenum to prevent brittleness.

2.1.4 Heat treatments

We learned earlier that although the iron atoms in austenite are more closely packed than in ferrite there are larger spaces between them. A result of this is that carbon is more soluble in austenite than in ferrite which means that carbon is taken into solution when steel is heated to a temperature at which the face-centred lattice exists. If this solution is rapidly cooled, i.e. quenched, the carbon is retained in solid solution and the steel transforms by a shearing mechanism to a strong hard microstructure called martensite. The higher the carbon content the lower is the cooling rate which will cause this transformation and, as a corollary, the higher the carbon content the harder will be the microstructure for the same cooling rate. This martensite is not as tough as ferrite and can be more susceptible to some forms of corrosion and cracking. We shall see in Chapter 11 that this is most important in considering the welding of steel. The readiness of a steel to form a hard microstructure is known as its hardenability which is a most important concept in welding. If martensite is formed by quenching and is then heated to an intermediate temperature (tempered), although it is softened, a proportion of its strength is retained with a substantial increase in toughness and ductility. Quenching and tempering are used to achieve the desired balance between strength, hardness and toughness of steels for various applications. If the austenite is cooled slowly in the first place the carbon cannot remain in solution and some is precipitated as iron carbide amongst the ferrite within a metallurgical structure called pearlite. The resulting structure can be seen under the microscope as a mixture of ferrite and pearlite grains.

With the addition of other alloying elements these mechanisms become extremely complicated, each element having its own effect on the transformation and, in particular, on the hardness. To allow the welding engineer to design welding procedures for a range of steels in a simple way formulae have been devised which enable the effect of the different alloying elements on hardenability to be allowed for in terms of their equivalent effect to that of carbon. One such commonly used formula is the IIW formula which gives the carbon equivalent of a steel in the carbon–manganese family as:

$$C_{eq} = C + \frac{Mn}{6} + \frac{Cr + Mo + V}{5} + \frac{Ni + Cu}{15} \qquad [2.1]$$

This represents percentage quantities by weight and what this formula says in effect is that weight for weight manganese has one-sixth of the hardening effect of carbon, chromium one-fifth and nickel one-fifteenth. This is a very scientific looking formula but it was derived from experimental observations, and perhaps one day someone will be able to show that it represents certain fundamentals in transformation mechanics. A typical maximum figure for the carbon equivalent which can be tolerated using conventional arc welding techniques without risking high heat affected zone hardness and hydrogen cracking is about 0.45%. Some fabrication specifications put an upper limit for heat affected zone hardness of 350 Hv to avoid hydrogen cracking but this is very arbitrary and depends on a range of circumstances. Limits are also placed on hardness to avoid stress corrosion cracking which can arise in some industrial applications such as pipelines carrying 'sour' gas, i.e. gas containing hydrogen sulphide.

The heat affected zone hardness can be limited by preheating which makes the parts warm or hot when welding starts and so reduces the rate at which the heat affected zone cools after welding. Preheat temperatures can be between 50° and 200°C depending on the hydrogen content of the welding consumable, the steel composition, the thickness and the welding heat input. For some hardenable steels in thick sections when the heat affected zone hardness remains high even with preheat, the level of retained hydrogen, and so the risk of cracking, can be reduced by post heating, i.e. maintaining the preheat temperature for some hours after welding.

Sometimes letting the work cool down slowly under fireproof blankets is sufficient. Where the composition, thickness or access makes preheating impracticable or ineffective an austenitic welding consumable can be used. This absorbs hydrogen instead of letting it concentrate in the heat affected zone but there is the disadvantage in that a hard heat affected zone still remains which may be susceptible to stress corrosion cracking; in addition the very different chemical compositions of the parent and weld metals may be unsuitable in certain environments.

2.1.5 Steels as engineering materials

Steels are used extensively in engineering products for a number of reasons. Firstly, the raw materials are abundant – iron is second only to aluminium in occurrence in the earth's crust but aluminium is much more costly to extract from its ore; secondly, steel making processes are relatively straightforward and for some types production can be augmented by re-cycling scrap steel; thirdly, many steels are readily formed and fabricated. The ability of carbon steels – in the welding context this means those steels with from 0·1% to 0·3% carbon – to have their properties changed by work hardening, heat treatment or alloying is of immense value. Perhaps the only downside to the carbon steels is their propensity to rust when exposed to air and water. The stainless steels are basically iron with 18–25% chromium, some also with nickel, and very little carbon. There are many types of stainless steel and care must be exercised in specifying them and in designing welding procedures to ensure that the chromium does not combine with carbon to form chromium carbide under the heat of welding. This combination depletes the chromium local to the weld and can lead to local loss of corrosion resistance. This can be seen in some old table knives where the blade has been welded to the tang and shows up as a line of pits near the bottom of the blade which is sometimes called 'weld decay'. To reduce the risk of this depletion of chromium the level of carbon can be reduced or titanium or niobium can be added; the carbon then combines with the titanium or the niobium in preference to the chromium. The most commonly known members of this family are the austenitic stainless steels in which nickel is introduced to keep the austenitic micro-structure in place at room temperature. They do not rust or stain when used for domestic purposes such as cooking, as does mild steel, but they are susceptible to some forms of corrosion, for example when used in an environment containing chloride ions such as water systems. These austenitic stainless steels are very ductile but do not have the yield point characteristic of the carbon steels and they do not exhibit a step change in fracture toughness with temperature as do the carbon steels. Some varieties retain their strength to higher temperatures than the carbon steels. The ferritic stainless steels contain no nickel and so are cheaper. They are somewhat stronger than austenitic stainless steels but are not so readily deep drawn. Procedures for their welding require particular care to avoid inducing brittleness. There is a further family of the stainless steels known as duplex stainless steels which contain a mixture of ferritic and austenitic structures. They are stronger than the austenitic stainless steels, and more resistant to stress corrosion cracking and are commonly used in process plant.

Metals other than the steels have better properties for certain uses, e.g. copper and aluminium have exceptional thermal and electrical conductivity. Used extensively in aerospace applications, aluminium and magnesium alloys are very light; titanium has a particularly good strength to weight ratio maintained to higher temperatures than the aluminium alloys. Nickel and its alloys (some with iron), including some of the 'stainless' steels, can withstand high temperatures and corrosive environments and are used in furnaces, gas turbines and chemical plant. However the extraction of these metals from their ores requires complicated and costly processes by comparison with those for iron and they are not as easily recycled. No other series of alloys has the all round usefulness and availability of the carbon steels.

For structural uses carbon–manganese steels have a largely unappreciated feature in their plastic behaviour. This not only facilitates a simple method of fabrication by cold forming but also offers the opportunity of economic structural design though the use of the 'plastic theory' described in Chapter 8. Whilst it may not be a fundamental drawback to their use, cognisance has to be taken of the fracture toughness transition with temperature in carbon steels.

2.1.6   Steel quality

The commercial economics and practicality of making steels leads to a variety of qualities of steel. Quality as used in this context refers to features which affect the weldability of the steel through composition and uniformity of consistency and the extent to which it is free from types of non-metallic constituents. The ordinary steelmaking processes deliver a mixture of steel with residues of the process comprising non-metallic slag. When this is cast into an ingot the steel solidifies first leaving a core of molten slag which eventually solidifies as the temperature of the ingot drops as in Fig. 2.3. Obviously this slag is not wanted and the top of the ingot is burned off. Since the steel maker doesn't want to discard any more steel than he has to, this cutting may err on the side of caution, in the cost sense, sometimes leaving some pieces of slag still hidden in the ingot. When the ingot is finally forged into a slab and then rolled this slag will become either a single layer within the plate, a lamination, or may break up into small

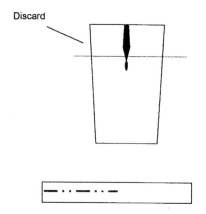

2.3 Formation of inclusions in a plate rolled from an ingot.

pieces called inclusions. For some uses of the steel these laminations or inclusions may be of no significance. For other uses such features may be undesirable because they represent potential weaknesses in the steel, they may give defective welded joints (Chapter 11) or they may obscure the steel or welds on it from effective examination by radiography and ultrasonics. In some steelmaking practices alloying elements may be added to the molten ingot but if they are not thoroughly mixed in these elements may tend to stay in the centre of the ingot, a plate rolled from which will have a layer of these elements concentrated along the middle of the plate thickness. Such segregation may also occur in steel made by the continuous casting process in which instead of being poured into a mould to make an ingot the steel is passed through a rectangular shaped aperture and progressively cooled as a continuous bar or slab. There are techniques for making steel more uniform by stirring before it is cast; non-metallic substances can be reduced by re-melting the steel in a vacuum or by adding elements which combine with non-metallic inclusions, which are mainly sulphides, to cause them to have round shapes rather than remain in a lamellar form. Such techniques obviously cost money and the steelmaker, as in other matters, has to strike a balance between cost and performance. Many of these steelmaking improvements were introduced initially in the 1960s and were extended in the early 1970s mainly as a result of the demands of the North Sea offshore oilfields development. As a result the quality of a large proportion of the world's structural steel production improved markedly and the expectations were reflected in the onshore construction industry.

Other developments in steelmaking practice were introduced in the following years aimed principally at improving the strength without detracting from the weldability or conversely to improve weldability without reducing the strength. These developments were in what were called the thermomechanical treatment of steel. Basically this comprised the rolling of the steel through a series of strictly controlled temperature ranges which modified the grain structure in a controlled way. As a result steels of fairly low carbon content, 'lean' steels, could be made with a strength which could formerly be obtained only by adding larger amounts of carbon. These developments created a confidence in the supply of conventional structural steel which became a relatively consistent and weldable product. However this position was not universal and even in the mid-1990s steel was still being made with what were, by then, old fashioned methods and whose consistency did not always meet what had become customary qualities. Certainly they met the standard specifications in composition but these standards had been compiled assuming that modern steelmaking methods were the only ones used. In one example the result was that although the steel had been analysed by conventional sampling methods and its composition shown to conform to the standard, the composition was not uniform through a plate. Virtually all the iron and carbon was on the outside of the plate with the de-oxidising and alloying elements in a band in the middle plane of the plate. Another example had bands in which carbon was concentrated which led to hydrogen cracking after gas cutting. The consequence of this is that precautions still have to be taken in design and fabrication to prevent the weaknesses of steel from damaging the integrity of the product. The most effective action is, of course, to ensure that the steel specification represents what the job needs. The question of cost or price is frequently raised but although steels of certain specification grades may cost more it is not because they are any different from the run of the mill production, it is that more testing, identification and documentation is demanded.

2.1.7 Steel specifications

An engineer who wants steel which can be fabricated in a certain way and which will perform the required duty needs to ensure that he prepares or calls up a specification which will meet his requirements. Most standard structural steel specifications represent what steelmakers can make and want to sell; anything beyond the basic product and any assurance level beyond that of the basic standard then requires an appeal to 'options' in the standard. The steel 'grade' is only a label for the composition of the steel as seen by the steelmaker and perhaps the welding engineer. It is not an identification for the benefit of the designer because the strength is influenced by the subsequent processing such as rolling into plates or sections. As a result the same 'grade' of steel may have widely differing strengths in different thicknesses because in rolling steel of the same composition down to smaller thicknesses its grain structure is altered and the strength is increased. As an example a typical structural steel plate specification calls for a minimum yield strength of 235 N/mm^2 in a 16 mm thickness but only 175 N/mm^2 in a 200 mm thickness, a 25% difference in strength. However it is not unusual for steels to have properties well in excess of the specified minimum, especially in the thinner plates. Whilst this may be satisfactory if strength is the only design criterion, such steel will be unsuitable for any structure which relies for its performance on plastic hinges or shakedown. The steel specification for this application must show the limits between which the yield strength must lie. Grades may be sub-divided into sub-grades, sometimes called 'qualities' with different fracture toughness properties, usually expressed as Charpy test results at various temperatures. Further, standard specifications exist to indicate the degree of freedom from laminations or inclusions by specifying the areas of such features, found by ultrasonic testing, which may be allowed in a certain area of the plate.

2.1.8 Weld metals

Weld metal is the metal in a welded joint which has been molten in the welding process and then solidified. It is usually a mixture of any filler metal and the parent metal, as well as any additions from the flux in the consumables, and will have an as-cast metallurgical structure. This structure will not be uniform because it will be diluted with more parent metal in weld runs, or passes, near the fusion boundary than away from it. This cast structure and the thermal history requires the consumable manufacturer to devise compositions which will, as far as is possible, replicate or match the properties of the wrought parent metal but in a cast metal. This can mean that the composition of the weld metal cannot be the same as the parent metal which in some environments can present a differential corrosion problem. As well as strength an important property to develop in the weld metal is ductility and notch toughness. Weld metals can be obtained to match the properties of most of the parent metals with which they are to be used.

2.2   Aluminium alloys

Aluminium is the third most common element in the Earth's crust after silicon and oxygen. The range of uses of aluminium and its alloys is surprisingly wide and includes cooking utensils, food packaging, beer kegs, heat exchangers, electrical cables, vehicle bodies and ship and aircraft structures. Pure aluminium is soft, resistant to many forms of corrosion, a good thermal and electrical conductor and readily welded. Alloys of aluminium variously with zinc, magnesium and copper are stronger and more suitable for structural purposes than the pure metal. Of these alloys, three series are suitable for arc welding; those with magnesium and silicon and those with magnesium and zinc can be strengthened by heat treatment and those with magnesium and manganese can be strengthened by cold working. Welding may reduce the strength in the region of the weld and in some alloys this strength is regained by natural ageing. In others, strength can be regained by a heat treatment, the feasibility of which will depend on the size of the fabrication. Allowances which have to be made for this loss of strength are given in design or application standards. A fourth series of alloys, aluminium–copper alloys, have good resistance to crack propagation and are used mainly for parts of airframes which operate usually in tension. In sheet form, this series is usually clad with a thin layer of pure aluminium on each side to prevent general corrosion; in greater thicknesses which may be machined they have to be painted to resist corrosion. These aluminium–copper alloys are unsuited to arc welding but the recently developed stir friction welding process offers a viable welding method. A valuable feature of aluminium alloys is their ability to be extruded so that complicated sections can be produced with simple and cheap tooling which also makes short runs of a section economical.

There is an international classification system for aluminium alloys summarised in Table 2.1. The system uses groups of four digits, the first digit giving the major grouping based on the principal alloying elements; the other digits refer to other features such as composition. Additional figures and letters may be added to indicate heat treatment conditions. The material published by the European Aluminium Association[3] is an authoritative source of knowledge about aluminium and its alloys.

Table 2.1 Summary of international aluminium alloy classification[3]

Alloy group series	Major alloying elements	Properties or uses
1 xxx	None	99% Al corrosion resistant
2 xxx	Cu	High strength, aerospace
3 xxx	Mn	Suitable for brazing
4 xxx	Si	Castings and filler wire
5 xxx	Mg	Medium strength
6 xxx	Mg + Si	Heat treatable
7 xxx	Zn + Mg	High strength, heat treatable
8 xxx	Other e.g. Sc, Li, Fe	

Fabrication processses

3.1 Origins

This chapter describes the principal features of the welding processes applied to those materials which are most commonly used in structural, mechanical and process plant engineering namely steels and aluminium alloys. To start with we need to be clear about what welding is in context of this book. Welding here is the joining of two or more pieces of metal so that the parts to be joined merge with one another forming a homogeneous whole across the connection. The word homogeneous is used guardedly here because although to the eye a weld may appear to be homogeneous, on a microscopic scale it may contain a range of different metallurgical structures and variations in the basic composition. It will be understood that this definition excludes soldering, brazing and adhesive bonding because joints made with those processes rely for the bond on an intermediate layer of a substance totally different from that being joined. Welding a metal requires the introduction of energy which can be as heat directly or in a form which will convert to heat where it is required. The earliest welding process, dating back thousands of years, was forge welding as applied to wrought iron where the parts to be joined are heated in a fire to a soft state and then hammered together so that one merges with the other. This is a traditional blacksmith's skill and it is most conveniently used for joining the scarfed ends of bars but it was used in joining the edges of strips to make gun barrels (Chapter 8). The modern analogue of this welding method is friction welding which will be referred to later on. Most other forms of welding involve melting the parts where they are to be joined so that they fuse together. This melting requires a heat source which can be directed at the area of the joint and moved along it. Such sources are the oxy-fuel gas flame and the electric arc. The flame or the arc can be used to melt the parts only (autogenous welding) but it is common to add filler metal of the same general nature as the metal being joined. Electric arc welding emerged towards the end of the nineteenth century and still represents the basis of a large proportion of all welding processes. Initially, in 1881, an arc from non-consumable carbon electrodes was used by August de Meritens and was patented by Benardos and Olszewski working in Paris. Shortly after that, in 1888, a Russian, N G Slavianoff, used a consumable bare steel rod as an electrode and he is generally accepted as the inventor of metal-arc welding. Bare wire electric arc welding was still in industrial use in 1935 and the author saw it still in use in 1955 for amateur car body restoration. The Swede, Oskar Kjellberg, patented the use of fusible coatings on electrodes in about 1910. However welding was slow to be taken up as an industrial process in heavy industry until the 1930s when it became applied on an industrial scale to ships, buildings and bridges. Even then the adoption of welding was not widely accepted until the Second World War gave urgency to many applications. Variations on the arc welding process blossomed, the individual bare or covered rod being followed by continuous electrodes, with and without coatings, which offered the opportunity of mechanisation. Submerged arc welding was introduced in the 1930s in both the USA and USSR as another means of continuous welding with the added benefits of an enclosed arc and in which the flux and wire combination could be varied to suit the requirements of the work. The principle of gas shielded welding was proposed in 1919 with a variety of gases being considered. In the 1930s attention concentrated on the inert gases but it was not until 1940 that experiments began in the USA using helium. Initially developed with a non-consumable tungsten electrode for the welding of aluminium the principle was to be applied to a continuous consumable electrode wire in 1948. This eventually led to the welding of steels in the 1960s on a production basis in the USA, UK and USSR by the development of techniques for using carbon dioxide as a shielding gas in place of the costly inert gases. Variations on this type of welding process came to be used in the form of wire with a core of flux or alloying metals and also wires with a core of a material which gave off carbon dioxide, fluorides or metal vapours thereby avoiding the need for a separate gas shield.

In the early 1960s attention was turned to the use of beams of energy in the form of electrons as a heat source for welding. Their effective use required operation in a vacuum and equipment and techniques soon followed which gave benefits in accuracy and precision with freedom from distortion and with metallurgical changes limited to a narrow band on either side of the weld. Ways of avoiding the disadvantages of *in vacuo* welding by techniques using partial vacuums are still being developed and no doubt will find applications in specialised markets. The constraints of vacuums were eventually circumscribed by the adoption of the laser beam as a heat source with the additional properties of being able to be transmitted around corners and of being capable of being split. The laser and electron beam processes today exist as complementary methods each being developed for the particular features which they offer.

At the same time as the esoteric high energy density beam processes were being developed attention was being paid to the development of friction welding, a far more mundane and mechanical bludgeon of a process. One of its advantages is that it does not actually melt the metal and so some of the metallurgical effects of arc welding are avoided. It rapidly gained industrial favour as a mass production tool, also in a version known as inertia welding, in the motor industry both in engine components such as valves, and transmission items such as axle casings; today, variations on the theme are still being invented and put to use. The latest is friction stir welding which amongst other uses has at last offered a metallic joining process with a potential for welding the aluminium–copper alloys commonly used in airframes because of their benign crack growth properties and absence of stress corrosion cracking in the atmospheric environment.

Another family of welding processes is the electrical resistance welding processes; in these the parts are clamped together between electrodes whilst an electric current is passed through them. The electrical resistance offered by the interface between the parts converts some of the electrical energy to heat which melts the interface and forms a weld nugget. This basic principle finds extensive use as *spot welding* in sheet metal fabrication in car bodies, white goods and similar applications and *seam welding* in more specialised fields. Trials of resistance spot welding of larger thicknesses of structural steels (~ 25 mm) were undertaken in France in the 1960s but did not lead to a practical method of fabrication. In contrast flash butt welding, another form of resistance welding, was extensively used in a range of thicknesses which amongst others found application in pipes and pipelines, particularly in the former Soviet Union. The parts are connected to an electrical power source and brought together and parted a number of times, on each occasion causing local arcing and melting until the whole interface is heated at which point the parts are forced together to make the final joint. The process is also used for joining as-rolled lengths of railway lines. On-site joining of the long lengths of line so manufactured continues to be one of the few applications of the thermit welding process. Basically an *in situ* chemical reaction between aluminium powder and iron oxide, it casts a pool of molten steel in the joint without the requirement for extraneous power supplies; it can be seen as an entertainment by night owls in cities all over the world which have tramlines.

Whilst mentioning the casting of pools of molten steel, the electroslag process is used as a means of joining thick sections of structural steel in one pass as in-line butts, tee-butts or cruciform joints. This can be faster than arc welding and less liable to give distortion; it can be performed in the vertical position only although its application can be extended to other positions by a version known as consumable guide welding. Variants of those processes mentioned above and other joining processes have been invented and either discarded along the way or left to serve a small specialised market.

A cynic might see arc welding as an extraordinary means by which to be joining materials in the twenty-first century. The material manufacturer produces a metal to fine limits of composition, microstructure and properties. Then it is subjected to a fierce arc so that the microstructure and properties of the metal adjacent to the weld are altered by the rapid heating and cooling. The process gives off toxic fume and, with the open arc processes, potentially injurious UV radiation. The resulting joint is erratic in shape, prone to fatigue cracking, possibly distorting the parts and with internal stresses much larger than any prudent designer would think of using. Arc welding has followed the pattern of other inventions which seem to be quite abominable but where the newcomers never seem to have the range of applications of the traditional ones. Perhaps it is that we get used to them, and the energy needed by human beings to change their habits and the money, time and effort invested in the traditional methods prevents or delays other means from emerging and themselves being developed. Another example of such inventions is the internal combustion piston engine as used in road vehicles. It has hundreds of moving parts being sent in one direction one moment and reversed the next, thousands of times a minute, scraping and hitting each other and wearing out. It can't start itself; it needs to be hand cranked or turned over with an electric motor which needs a huge battery, much larger than other services require, and so is just dead weight for the rest of the time. To allow the engine to keep running when it takes up the drive it has to have a slipping transmission, either a solid friction or hydraulic clutch, which wastes energy. The engine has such a small effective working speed range that it has to have a transmission which has to be manually or mechanically reconfigured in steps to keep the engine speed within the working range. It sends out noise and toxic gases and particles and the used lubricating oil is poisonous and environmentally damaging unless re-processed. It sounds like some Emmett cartoon machine; would we really start from there if we had to invent an engine today? Nonetheless taking the pragmatic view we now see highly developed arc welding processes which can make reliable joints giving a performance consistent with that of the parent metals.

3.2   Basic features of the commonly used welding processes

3.2.1   Manual metal arc welding

This process is what probably comes to most people's minds when arc

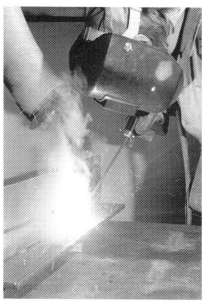

3.1 Manual metal arc welding with a covered electrode (photograph by courtesy of TWI).

welding is mentioned. The welder holds in a clamp, or holder, a length of steel wire, coated with a flux consisting of minerals, called a welding electrode or rod; the holder is connected to one pole of an electricity supply. The metal part to be welded is connected to the other pole of the supply and as the welder brings the tip of the rod close to it an arc starts between them (Fig. 3.1). The arc melts the part locally as well as melting off the end of the rod. The molten end of the rod is projected across the arc in a stream of droplets by magneto-electric forces. If the welder moves the rod along the surface of the part keeping its end the same distance from the surface a line of metal will be deposited which is fused with the molten surface of the part, forming weld metal, and will cool and solidify rapidly as the arc moves on. The flux coating of the electrode melts in the heat of the arc and vaporises so giving an atmosphere in which the arc remains stable and in which the molten metal is protected from the air which could oxidise it; the flux also takes part in metallurgical refining actions in the weld pool. Some types of flux also contain iron or other elements which melt into the weld metal to produce the required composition and properties. Rods for manual metal arc welding are made in a variety of diameters typically from 2·5 mm to 10 mm in lengths ranging between 200 mm and 450 mm. There are many different types of electrodes, even for the carbon–manganese steel family. The main differences between them lie in the flux coating. There are three main groups of coating in the electrodes used in most conventional fabrications.

- *Rutile* coatings include a high proportion of titanium oxide. Rods with this type of coating are relatively easy to use and might be called general purpose rods for jobs where close control of mechanical properties is not required. The steels on which they are used should have good weldability. In practice this means mild steel.
- *Basic coatings* contain lime (calcium carbonate) and fluorspar (calcium fluoride). They produce weld metal for work where higher strength than mild steel is required and where fracture toughness has to be controlled. They are used where the level of hydrogen has to be controlled as in the case of more hardenable steels to prevent heat affected zone hydrogen cracking. Rods with this type of coating are more difficult to use than those with rutile coatings, the arc is more difficult to control and an even weld surface profile more difficult to produce. The need for low hydrogen levels means that they may be sold in hermetically sealed packs; if not, they must be baked in an oven at a specified temperature and time and then kept in heated containers, or quivers, until each is taken for immediate use.
- *Cellulosic* coatings have a high proportion of combustible organic materials in them to produce a fierce penetrating arc and are often used in the root run in pipeline welding, 'stovepipe welding' as it is called, and for the capping run. The high quantities of hydrogen which are released from the coating require that precautions be taken to prevent hydrogen cracking in the steel after welding.

Rutile and basic coated rods may have iron powder added to the coating. This increases productivity by producing more weld metal for the same size of core wire. The larger weld pool which is created means that iron powder rods cannot be as readily used in all positions as the plain rod. Covered electrodes are also available for welding stainless steels and nickel alloys but are proportionately less popular than for carbon steels; much of the work on these alloys is done with gas shielded welding. The electrical power source for this type of welding can be a transformer working off the mains or an engine driven generator for site work. The supply can be AC or DC depending on the type of rod and local practice.

3.2.2 Submerged arc welding

This process uses a continuous bare wire electrode and a separate flux added over the joint separately in the form of granules or powder. The arc is completely enclosed by the flux so that a high current can be used without the risk of air entrainment or severe spatter but otherwise the flux performs the same functions as the flux in manual metal arc welding (Fig. 3.2). At high currents the weld pool has a deep penetration into the parent metal and

3.2   Submerged arc welding (photograph by courtesy of TWI).

thicker sections can be welded without edge preparation than with manual metal arc welding. Lower currents can of course be used and with the ability to vary welding speed as well as the flux and wire combinations the welding engineer can achieve any required welded joint properties. The process has the safety benefit of there not being a continuously visible arc.

The process is most commonly used in a mechanised system feeding a continuous length of wire from a coil on a tractor unit which carries the welding head along the joint or on a fixed head with the work traversed or rotated under it. When welding steels a welding head may feed several wires, one behind another. Both AC or DC can be used and with a multi-head unit DC and AC may be used on the different wires; DC on the leading wire will give deep penetration and AC on the other wires will provide a high weld metal deposition rate. Welding currents of up to 1 000 A per wire can be used. Manually operated versions of submerged arc welding are used in which the current levels are limited to some 400 A.

The fluxes used in submerged arc welding of steels can be classified by their method of manufacture and their chemical characteristics. They may be made by melting their constituents together and then grinding the solidified mix when it has cooled, or by bonding the constituents together into granular form. The chemical characteristics range from the acid types containing manganese or calcium silicates together with silica to the basic types, again containing calcium silicates usually with alumina, but with a lower proportion of silica than the acid types. The acid fluxes are used for general purpose work whereas the basic fluxes are used for welds requiring control of fracture toughness and for steels of high hardenability to avoid hydrogen cracking.

The wire is usually of a 0·1% carbon steel with a manganese content of between 0·5% and 2% with a relatively low silicon content around 0·2%. As a mechanical process, submerged arc welding is capable of greater consistency and productivity than manual welding although to balance this the process is not suited to areas of difficult access and multi-position work *in situ.*

3.2.3 Gas shielded welding

3.2.3.1 *Consumable electrodes*

Here a bare wire electrode is used, as with submerged arc, but a gas is fed around the arc and the weld pool (Fig. 3.3). As does the flux in the manual metal arc and submerged arc processes this gas prevents contamination of the wire and weld pool by air and provides an atmosphere in which a stable arc will operate. The gas used is one of the inert gases, helium or argon, for non-ferrous metals such as aluminium, titanium and nickel alloys, when the process is called metal inert gas (MIG). For carbon steels pure carbon dioxide (CO_2) or a mixture of it with argon is used when the process is called metal active gas (MAG). The functions of the flux in the other processes have to be implemented through the use of a wire containing de-oxidising elements, about 1% manganese and 1% silicon. These combine with the 'active', i.e. the oxygen, part of the shielding gas and protect the molten steel from chemical reactions which would cause porosity in the weld. For stainless steels a mixture of argon and oxygen may be used.

The range of currents which can be used covers that of both the manual metal arc and the lower ranges of the submerged arc processes. The wire is fed from a coil to a welding head or gun which may be hand held or mounted on a mechanised system. The wire may be solid or it may have a core containing a flux or metal powder which gives the ability to vary the weld metal properties by choice of the wire. The need for gas and wire feed conduits and, in the case of higher currents, cooling water tubes, can make the process rather more cumbersome to use than manual metal arc and restricts its application in site work. The variation of the process, self shielded welding, in which the core is filled with a chemical which emits shielding vapours on heating eliminates the need for a gas supply and is used

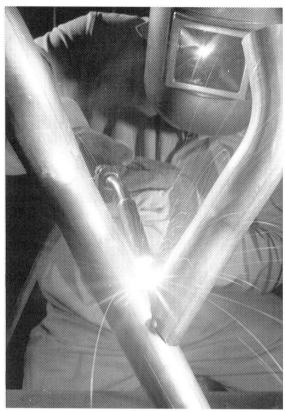

3.3 Gas shielded welding (photograph by courtesy of TWI).

satisfactorily on site. The solid wire gas shielded process has the advantage in production work over the flux processes in that the welds do not need as much de-slagging, but small 'islands' of silicates may remain on the weld surface and have to be removed if a paint system is to be applied. A flux process with a self releasing slag will have the advantage over solid wire where the weld has to be brushed.

DC is used in one of two modes. At low currents the transfer of metal from the wire to the weld pool takes place after short circuits as the tip of the wire intermittently touches the weld pool. This is called dip transfer. At high currents the transfer is by a stream of droplets propelled across the arc and termed spray transfer. The dip transfer mode is used for sheet metal work, root runs and for positional work, i.e. overhead or vertical welds. Except with rutile flux cored wires, the spray transfer mode is unsuited to positional welding and is used for downhand filling runs in thicker material where the greater deposition rate can be employed with advantage. A wider control of metal transfer can be achieved by pulsing the welding current using a special purpose power source. This permits a wider range of conditions for positional welding but cannot be used with pure carbon dioxide as a shielding gas. It is restricted to welding with argon–CO_2 –oxygen mixtures.

3.2.3.2 Non-consumable electrodes

For thin sheet work and precision welding of components to close tolerances the tungsten inert gas (TIG) process can be used. The arc is struck between a tungsten electrode and the workpiece with argon or helium as the shielding gas. The tungsten electrode is not consumed and filler can be added to the weld as a wire although many applications employ a joint design in which a filler is not required (autogenous welding) (Fig. 3.4). AC is used for aluminium alloys and DC for ferrous materials. The TIG process can be used manually or mechanised. A process with similar applications at low currents is the microplasma process. A jet of plasma is produced in a torch which looks similar externally to a TIG torch. It can be used for very fine work on a variety of metals. The plasma process used at high currents, e.g. 400 A, can be used for butt welding; the mechanism here is different from TIG and microplasma. The plasma jet melts through the metal and forms a hole in the shape of a keyhole; as the torch moves along the joint the metal re-solidifies behind the keyhole so as to fuse the two parts. The process is

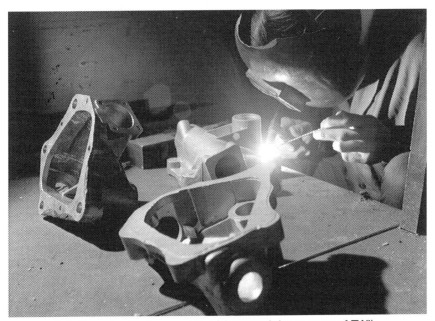

3.4 Tungsten inert gas welding (photograph by courtesy of TWI).

used in a mechanised form for welding stainless steel and aluminium alloys, and is particularly suited to pipe and tubular shapes in which the joint can be rotated under a fixed welding head.

3.3   Cutting

Structural steels are usually gas cut although laser cutting is increasingly used for plate. In gas cutting a flame of fuel gas such as acetylene burning in oxygen heats the area to be cut; a stream of oxygen is then injected around the flame which actually burns the steel and ejects the oxide as dross. The cutting torch may be hand held or it may be mounted on a mechanised carriage. Depending on the thickness the steel has to be pre-heated as for welding to prevent a hard heat affected zone being formed on the cut edge with the attendant risk of cracking. A cutting procedure specification can be prepared and tested in a manner analogous to a welding procedure specification. Mechanised cutting is preferred as it can produce a smoother edge than manual cutting; the burners can be traversed in two directions to cut shapes or holes. Numbers of cutting heads can be used simultaneously so that many copies of the same shape can be cut. It goes without saying that computer control can be applied as a first phase of a computer aided manufacturing system. The cutting head can be set at an angle so that a bevelled edge can be cut as a weld edge preparation. Two, or even three, heads can be mounted as shown in Chapter 4 so that a double bevel with a root face can be cut in one pass. A properly adjusted gas cutter will leave a smooth edge although inclusions or laminations in steel plates can blow out gases leaving a local roughness in the cut. The cut may carry a glaze of silicates from the steel which may prevent paint adhering to the surface. For this reason it is usual to grind or grit blast the surface if it is to be painted.

Thin sheet and plate metals (< 12 mm) can be cut by guillotine and holes punched. In both cases the sheared edge is severely cold strained and in some carbon steels may offer poor fracture toughness if welded; a welding procedure test should be able to clarify this point. Stainless steels do not cut well by burning and are frequently plasma cut. Mechanised versions of plasma cutting equipment can make the cut underwater which gives a very clean cut with little distortion. Carbon and stainless steels can be cut very accurately with lasers up to thicknesses of 20 mm.

3.4   Bending

Steels can be bent cold, i.e. at ambient temperature, although they have to be formulated to be able to do this uniformly in complex shapes such as car bodies. The steels used for cold formed building shapes such as hollow sections, purlins, rails and floor beams are similar to the carbon–manganese steels used in rolled sections; perhaps the oldest example of a cold formed section is the well known corrugated iron to be found all over the world. There is a limit to the thickness of steel which can be cold formed, partly owing to the mechanical force required and also because above a certain thickness the plastic straining can leave a material which can be, particularly on welding, of a reduced fracture toughness. This fracture toughness can be recovered by a heat treatment similar to thermal stress relieving.

The alternative is to bend or roll the steel at a high temperature such as that used for stress relief. There is no suitable intermediate temperature for forming because brittleness will be induced. Stainless steels can be formed cold but they have a larger springback than carbon steels. Aluminium and its alloys can be folded or formed in various ways and have a particular advantage in that they can be extruded through a simple die to generate an almost infinite variety of profiles.

3.5   Residual stresses and distortion

The progression of the arc along the joint represents a moving heat source forming a pool of molten metal around which is a complicated and changing distribution of temperature and strain as the metal is progressively heated, melted, frozen and cooled. In a multi-run joint the complexity is compounded by the presence of the earlier runs. One can illustrate the formation of residual stresses by a simplified model of a butt weld which ignores the progressive aspect of welding. Fig. 3.5 shows two plates and between them a hot strip of metal representing the weld area. If the weld were free to contract lengthwise on cooling it would end up being shorter

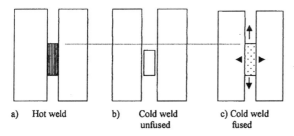

a)   Hot weld   b)   Cold weld   c) Cold weld
                              unfused   fused

3.5   Origins of residual stresses in welded joints.

than its hot length. By being fused to the plates it is restrained by them and so is put into tension; conversely the plate is put into compression. The distribution of stress is more complicated than the simple model would show and measurements suggest that it would be of the general form as in Fig. 3.6. In the direction transverse to the weld the simple model would not produce any residual stress except that from the Poisson effect. In practice of course

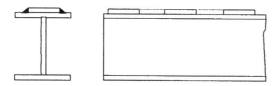

3.6 Typical distribution of residual stresses in a butt welded plate.

the progressive nature of the weld means that the start is cool and able to react forces quite soon after the heat source moves away; the result of this is that transverse residual stress system is established also shown in Fig. 3.6.

Residual stresses are the main source of distortion. They represent a self-equilibrating stress system, that is a set of stresses which are reacted entirely within the item in question. In the simplest case a bar made from two pieces welded end to end will shrink from its hot length when it cools down as the weld and adjacent metal contract. A weld on the surface of a plate will make it bend out of its plane and an item welded to a beam flange will tend to bend the beam by the contraction of the weld and adjacent metal. Even unwelded rolled sections contain residual stresses because the different thicknesses cool at different rates after the section is rolled. The effects can be seen if an I section is slit to make two T sections when they will bow as a result of the residual stresses which were balanced within the complete section. When a cover plate is to be welded to a beam flange (Fig. 3.7) it

3.7 Reduction of distortion by welding intermittent pads instead of continuous plate.

must be welded symmetrically otherwise the beam will curve in the plane of the cover plate. Even so it may cause the beam to curve in the other plane. This distortion can be minimised by stitch welding, that is by making relatively short lengths of weld at intervals and filling in between them. If the cover plate does not really need to be continuous the distortion can be minimised by cutting it into shorter lengths. This approach is particularly suitable when the plate is really just a mounting block for equipment.

Residual stresses and distortion can affect the load bearing capacity of a structure by reducing the buckling strength of a member; this is dealt with in Chapter 8.

3.6   Post weld heat treatment

Steel fabrications may be subjected to post weld heat treatment (PWHT). This usually refers to a treatment in an oven or with external electrical or chemical heating in which the fabrication is heated to between 580° and 620°C, a bright red heat, for about 1 h for each 25 mm of thickness. The most common reason for doing this is to relax the residual stresses set up by welding. This has two potential benefits: one is to stabilise the fabrication against distortion in machining or service and the other is to decrease the risk of brittle fracture. The latter effect is also enhanced by the effect of the heating on the microstructure local to any potential fracture initiation sites.

The heat treatable aluminium alloys may be heat treated to artificially age the welds and regain the strength of the parent metal.

Considerations in designing a welded joint

4.1 Joints and welds

It is convenient to define a *joint* separately from a *weld*. The joint is the manner in which the parts meet each other, e.g. butt joint, lap joint, T joint, corner joint, as shown in Fig. 4.1. A butt joint is where two parts butt against each other end to end or edge to edge, whereas a lap joint is where the two parts overlap. A T joint is so called because the parts, if they are of simple shape such as flat plate, meet in the form of a T. A corner joint is so called

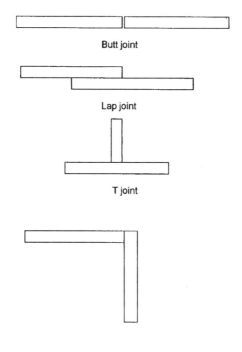

Butt joint

Lap joint

T joint

Corner joint

4.1 Joint forms.

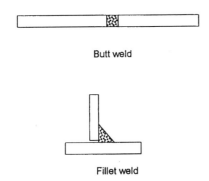

Butt weld

Fillet weld

4.2 Basic weld types.

because it forms an angle, or corner, where the two parts meet. Welds used to make these joints with arc welding can conveniently be thought of as of only two main types, the butt weld and the fillet weld, as in Fig. 4.2. The butt weld joins two pieces by fusing their complete cross sections so creating a monolithic object whereas the fillet weld connects the two parts with a line of weld metal without attempting to create a full section joint. Even so these weld names are somewhat arbitrary – the weld metal doesn't know whether it is in a butt weld or a fillet weld. The butt welded joint has a potential for a higher performance than a joint made with a fillet weld but it can be more costly to make. The butt weld is capable of being examined for internal soundness to give confidence that the weld will perform as required. The fillet weld gives a lower performance, cheaper, joint which will still provide a load carrying connection but which cannot be examined for internal soundness as readily as a butt weld. This suggests that in general one should not place as high a level of confidence in the fillet weld's performance as in that of a butt weld; this means that these two types of weld are not just different in form, they also represent two different engineering philosophies. Requirements for welder qualifications in many fabrication contracts would suggest that the manual skill required to make a butt weld is greater than that for a fillet weld which places a premium on the supply of welders for manual butt welds. It is difficult to understand why this is held to be so; the technique for achieving full root fusion in a fillet weld is very demanding and there have been occasions when qualified butt welders have failed a fillet weld test because of this very feature. Logic would also perhaps suggest that if fillet welds cannot be non destructively examined as effectively as butt welds then a fillet welder needs to be more skilled than a butt welder. As with many aspects of welding this attitude has probably grown out of tradition and not out of logic. The basis for this would be that since butt welds are generally used in high integrity applications and fillet welds are used in low integrity applications then welders qualifying for fillet welding need to be less skilled than those for butt welds. Just to confuse the issue

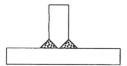

4.3 T joint made with butt and fillet welds.

further there can be butt welds with partial penetration (NB: *lack* of penetration is a defect, see Chapter 11) and types of joint made with a combination of butt and fillet welds, as in Fig. 4.3.

When we nominate welding as the joining method we have to choose a material which, when welded, will perform as required in service. The type of joint we can use is influenced, or even defined, by the nature of the object of which it is part. The choice of type of weld is then limited to one of the few which will satisfy the demands on the joint both in terms of service performance and accessibility for welding and inspection. The latter will depend on the chosen welding process; in practice the choice is narrow and most have to make do with the few processes actually available. Take as an example a simple joint connecting the edges of two steel plates of the same composition, of equal thickness and in the same plane. For reasons of structural performance we might opt for a full penetration weld. Theoretically the choice of welding processes is great, ranging from manual metal arc welding with coated electrodes to electron beam and laser; we might think also of electroslag, diffusion bonding, friction welding and flash butt welding. Much depends on the industry in which we are working, its traditions, its expectations and its manufacturing sophistication in terms of materials, dimensional tolerances, surface finish and cleanliness. We have to recognise the restraint of cost, the size and shape of the fabrication and whether or not it is to be mass produced or a one off. Unless we have a really pressing case for a high tech and expensive welding process we will end up with one of the more mundane processes. The choice will be further whittled down to the facilities of the fabricator. Most work still ends up being done with simple arc welding.

There are a number of other factors which will influence the choice of the joint and weld. A most important one is that of feasibility of inspection, for despite the best of intentions the ideal of on-line process control based on the qualities of the weld being made still evades much mechanised welding and of course has no role in manual welding. We therefore still find a lot of work being inspected after completion by various means ranging from visual surface examination, assisted visual examination such as magnetic particle and dye penetrant, to radiography and ultrasonics (methods described in Chapter 11) and the relatively more esoteric but well established techniques such as eddy currents and ultrasonic imaging. All of these techniques aim to discover physical discontinuities in the joint on a macro scale such as are represented by lack of fusion, cracks, porosity, inclusions and laminations. The methods all rely on detecting the boundary between solid metal and cavities and their success presupposes that no such cavities are intended to be there such as in partial penetration butt welds and fillet welded joints. This means that if full confidence in such inspection is required we have to use a full penetration butt weld; in addition we have to be sure that the internal structure of the steel does not itself contain cavities or inclusions on a macro or micro scale so distributed that they will confuse the inspection technique. There are techniques for internal examination of fillet welds but these are rather specialised and not in common use. All of these techniques have their individual requirements for access which have to be taken into account when designing the joint and weld.

It is not surprising that there can be conflict between these considerations and as in many other walks of life the designer of the welded joint has to make compromises. The necessity and scope for compromise is raised in other chapters of this book from which it will become apparent that as in other fields of engineering design there is no unique 'correct' solution although there may be a best or most expedient solution for a particular set of circumstances. Table 4.1 lists many of the considerations in designing a welded joint.

4.2   Terminology

This section might well be entitled 'Communication' for it is about the means by which instructions are conveyed between people. Spoken and written language is vital to most human endeavour and its mastery eludes most of us. Because of the history of the British Isles over the past three thousand years the English language is derived from the languages of the Celts, the Romans, the Angles, the Saxons, the Vikings and the Normans, who were themselves of Viking origin but over centuries had adopted the French culture. This gives the English language the ability to represent objects or ideas in many ways and with more nuances than many. Even in the twenty-first century some words have not travelled far from the locality in which their original users settled. For convenience and from frequency of use every trade and profession develops a narrower interpretation of some of the common words and invents some of its own.

The absorption of the various languages into the British Isles over the first millennium AD, and up to the time of the Normans, produced a conglomerate language which became known as English, most famously used by Chaucer in the fourteenth century and by Shakespeare in the sixteenth and seventeenth centuries. Their writings became available to more than the small group of educated people through the printing press invented in the fifteenth century by Gutenberg in Germany. The second half of the

Table 4.1. Considerations in designing a welded joint

Feature	Examples of matters for consideration
Service performance	static strength ductility fatigue life corrosion resistance

Material weldability	as-welded strength
	    " ductility
	    " fracture toughness
	chemical composition
	susceptibility to cracking

Welding consumables	matching parent metal properties
Welding process	weld type access material size of component cost shop or site
Distortion	weld preparation single or double sided weld heat input weld run sequence
Access for welding	position configuration reach obstruction shop or site
Access for inspection and NDT	as for welding

Cost	building and plant charges –
	   e.g. interest on capital, depreciation, leasing
	   charges, maintenance
	consumables, materials, energy
	payroll costs
	overheads
	taxes

Weld quality standard NDT methods	joint and weld configuration
Position of joint in fabrication	service stress size of component shop or site work transport access for welding/inspection

second millennium AD saw this language moved to other parts of the world by Britons who settled in other continents.

Words used in all walks of life may remain unchanged over centuries in one country whilst changing in another; for example, some usage of English words common in the USA is now seen as old fashioned in the UK. Their meanings were common in Britain in the seventeenth and eighteenth centuries when the early settlers crossed the Atlantic but have since passed into disuse in their country of origin. A similar position exists with French as spoken in Canada. An example in the English used in the USA is a 'chapter' in the sense of a group of people belonging to a larger organisation which in the UK is now called a branch except in some churches which are more resistant to change than most. In Australia 'manchester' is a word applied to cotton goods because at the time of the British settlement of Australia Manchester was the city at the centre of the cotton trade in Britain and the world; the word has since passed into disuse in the UK. We should not be surprised then if the English language terminology used in welding and welded joints can vary even within one country and the terms used for the same thing may differ even between industries in a country or between different groups of people in the same industry. *Welding rod* is a term well recognised on the shop floor and *welding electrode* is less colloquial whilst the formal written term might be *covered electrode* which few in industry would use in speech. For the sake of clarity in conveying instructions most countries establish a formal terminology by issuing standard vocabularies. Alongside these are international dictionaries which offer the equivalent words in a number of languages. The terminology given in this chapter is based on that commonly used in the UK which is published in BS 499.

In engineering much of the instruction is conveyed in the form of drawings in which, for simplicity, symbols are used in place of text. This helps to avoid ambiguity and in international trade also avoids the potential problems associated with having to translate text. Nonetheless there comes a point when a symbolic representation may become too fussy and confused at which time the draughtsman may resort to detail scrap views. There are standards at all levels giving symbols for use on drawings relating to the written terms; the international level is represented by ISO 2553. Figures 4.1–4.4 show the basic joints and weld terminology. There are a few more terms in common use which are needed to define a weld. There are national, regional and international standards which give terminology.[4–7]

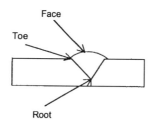

BUTT WELD

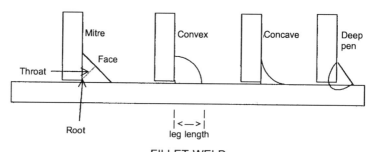

FILLET WELD

4.4 Commonly used weld terminology.

4.3   Weld preparations

4.3.1   In-line butt joints

With the welding conditions for rutile and basic low hydrogen electrodes used for most manual metal arc welding there is very little penetration at all. It follows then that when a butt weld is to be made between the edges of the plate they have to be bevelled so that the weld metal can be placed in the joint and fused with the parent metal. Cellulosic coated electrodes give a more widely penetrating arc (Fig. 4.5) and are used for root runs in some structural steelwork applications but more commonly in pipeline circumferential welds made on site using a technique called stovepipe welding. These electrodes release a higher level of hydrogen than the other two types and the welding procedures have to be designed to recognise this so as to avoid heat affected zone cracking.

With mechanised MAG or submerged arc welding equipment a full penetration butt weld can be made from one side without edge bevels if the welding current is high enough. However this requires edges having a close fit all the way along the joint and that the welding conditions are previously

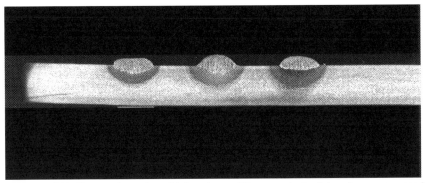

4.5 Penetration of weld bead on 10 mm steel plate; 4 mm covered electrodes at 167 A. Coating types from left to right: cellulosic, basic, rutile.

proven and controlled during the welding of the joint. If these matters are not attended to the arc will either fail to penetrate the thickness of the material leaving a lack of penetration defect or it will blow through in a cutting action which will not leave a fully fused joint. If occasional lengths of lack of penetration can be tolerated then this method can be used with the welding conditions set deliberately to offer lack of penetration rather than blow through; where full penetration is essential the root of the weld can be ground out or gouged from the opposite side (back gouged) and a weld made on that side. For relatively thick materials a middle path can be followed where the edge is bevelled over some of the depth of the joint and the first run is designed to penetrate the root face and subsequent runs are made in the preparation. The weld can be made with a run made successively from each side of the joint with the second run made over the as-welded root of the first run. In doing this there is a risk of sporadic lengths of lack of penetration or slag inclusions. As with single sided welds if a sound root in the finished weld is essential then the root of the first run must be ground or back gouged leaving a clean groove for the second weld, Fig. 4.6(a). The root face must be kept to a minimum depth otherwise a large amount of metal is left to be gouged out which is not only costly but results in the introduction of a great deal of heat and the risk of excessive distortion. When selecting a weld preparation distortion is one of the factors to be taken into account.

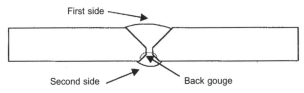

4.6 (a) Back gouging in a welding sequence.

With high welding currents the weld pool is large and surface tension effects are relatively less pronounced; the weld metal can then run out of the joint or may flow ahead of and under the arc preventing its striking the parent metal and creating a lack of fusion defect. As a result high welding currents are used only in the flat or the horizontal vertical position. Another point is that with high heat inputs the weld metal is virtually as cast with perhaps a large grain size which may have poor properties such as fracture toughness; when using a number of smaller, lower heat input runs (or passes) each run heat treats the previous run and improves the properties. For these reasons some application specifications restrict the heat input to, for example, 5 kJ/mm. As well as taking less time, few large runs have another advantage over a number of small ones in that angular distortion can be less. A compromise arises here depending on the demands of the specification.

The choice of the weld preparation is based on the configuration of the joint, the access for welding and inspection and the cost or the type of cutting equipment available. The simplest type of edge preparation is the single bevel, as shown in Fig. 4.6(b). This can be made very cheaply by gas cutting. The *root face* is left because the arc would melt a sharp edge, or feather edge, making a consistent weld difficult. In addition any wander in the cutting line, Fig. 4.6(c), does not change the position of the edge so maintaining the consistent root gap required for a sound root. About the only joint for which the feather edge is suitable is where the butt weld is made onto a backing strip or bar where the position of the plate edge is not so important and where the edge may be fused into the backing in any case.

As material thickness increases it may be desirable to change from a single sided bevel to a two sided bevel, Fig. 4.6(d), for two reasons. Firstly, the volume of weld metal and thereby the cost is reduced, Fig. 4.6(e) secondly, the heat input and thermal history is more balanced through the thickness, leading to lower levels of distortion. To minimise distortion the

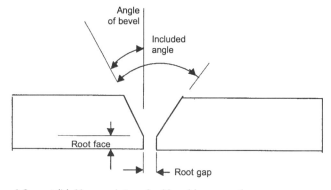

4.6 cont (b) Nomenclature for V weld preparation.

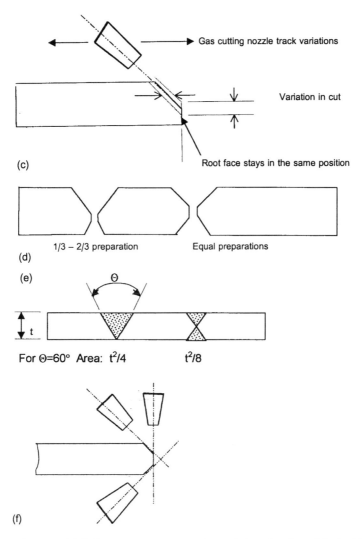

4.6 cont (c) Tolerance on gas cutting of bevel edge; (d) double V preparations; (e) relative volume of weld metal in weld preparations; (f) single pass cutting of a plate edge with double bevel.

preparation is not made symmetrical – the size of the preparation on the first side to be welded is less than that on the second side. The root face serves the same purpose as with the single sided preparation. The double sided bevel preparation can be gas cut quite quickly and cheaply in one pass with a three burner cutting head, Fig. 4.6(f).

More complicated edge preparations are used to reduce weld metal volume and distortion. These are based on cutting a curved rebate in the edge which can be either from one side only giving an in-line butt weld and

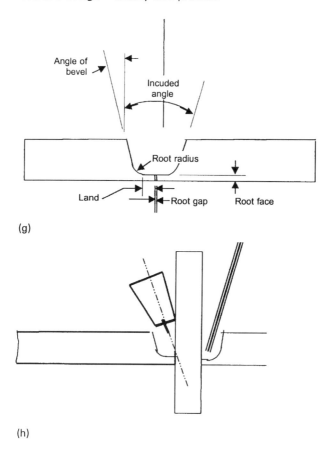

(g)

(h)

4.6 cont (g) Nomenclature for single U weld preparation; (h) limitation of access to MIG/MAG nozzle iin J preparation.

is called a U preparation, Fig. 4.6(g), or from both sides giving an symmetrical double U preparation. Where this preparation is on one plate only, as for example for a T joint, it is called a J preparation. These types of preparation are not readily gas cut, except by gouging the assembled joint, and the most common method of making them is by machining. This immediately adds another cost and a requirement for machine tools other than a gas cutting table. The U and J preparations introduce another consideration which is that of access for the welding arc, Fig. 4.6(h). As the thickness of the material increases, the cost and distortion benefits of the U and J preparations increase but there comes a depth of the preparation beyond which the manually held gas shielded welding torch, or gun, is too big for the space in which it has to be operated. The arc cannot be directed at the correct angle to the sidewalls or the root and there is a risk of lack of fusion defects. In these circumstances a manual metal arc electrode can still

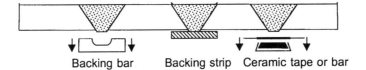

Backing bar   Backing strip   Ceramic tape or bar

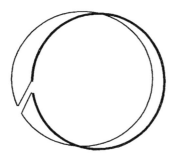

Sprung backing ring

4.7 Types of backing materials.

be used satisfactorily and to some extent a shielded cored wire gun which has a smaller diameter than one with a separate gas shield.

Making a butt weld from one side only as in a tube requires considerable skill. The making of such a weld can be eased by the use of backing strips or backing bars. Fig. 4.7 shows three types. The backing bar, which may be of copper or other good heat conductor, is not fused into the joint and is removed after welding. This device is used extensively in automatic pipe welding machines where the backing bar or back-up bar is attached to the power operated clamp which keeps the pipe ends in alignment during welding. The backing strip is made of the same metal as that being joined and remains in position; it is therefore not suitable for pipes carrying fluids and also has a relatively low fatigue performance (see Chapter 6). The ceramic backing, in the form of a bar or a tape, is used for fine work where a smooth internal profile is required and is removed after welding. The ceramic or plain backing can be made as a sprung ring which can be slid into the tube and will be a close fit to the weld root. A butt weld on a backing can be used to provide assembly tolerances. If, for example, the ends of a plate or tube cannot be positioned accurately, a butt weld on a backing will in effect offer a 'sliding joint'. On a plate it may be convenient to tack weld the backing strip to one or other side of the strip, as in Fig. 4.8(a). Tube ends can be machined to make a spigot joint, Fig. 4.8(b). On heavy wall tubes such as piles a backing strip can be attached to the stab-in guide, shown in Fig. 4.8(c).

(a) Attachment of backing strip showing alternative positions of tack welds

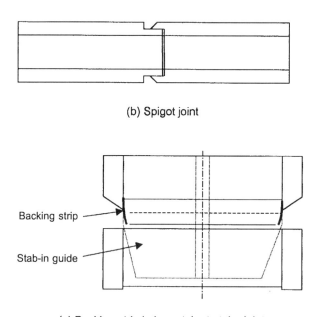

(b) Spigot joint

Backing strip

Stab-in guide

(c) Backing strip in large tube to tube joint

4.8 Types of weld backing.

Machining edge preparations on flat plates requires a planer or milling machine which may represent an expensive piece of capital equipment whereas the machining of preparations on tubes or circular bars requires a lathe which may be readily available in most machine shops. The choice of edge preparation may then depend not only on the welding requirements but on the shape of the parts being joined. Preparations made either by gas cutting or by machining will require attention before welding commences. Gas cut edges must be cleaned of scale and dross and any areas of erratic cut dressed smooth. Machined edges must be cleaned of oil. In both cases if the parts have been exposed to weather they may need to be cleaned of rust or any paint and preservative coatings. Whatever method is used to form an edge preparation the opportunity must be taken to examine the cut edge for evidence of laminations, inclusions or other defects in the material which could create defects in the weld or appear as weld defects in any non destructive examination of the weld. The process of gas cutting itself will

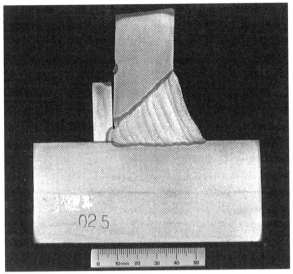

4.9   T butt weld on a backing strip.

sometimes reveal the presence of inclusions or laminations in plate. The heat from the cutting may expand the gases in inclusions which then disturb the cutting gas stream creating an irregular cut; even the sound of this happening will point out a potential problem to the experienced operator.

4.3.2   T joints

The edge preparation may be in the form of a single or double bevel or J preparation. Fig. 4.9 shows a T joint between a 50 mm and a 30 mm plate. This is a butt weld on a backing strip where the pattern of the weld runs show that it has been made in the vertical up position with the welding rod being weaved, i.e. moved from side to side. Other features to note are that the backing strip tack weld has been broken by the rotation of the upstand plate caused by the thermal shrinkage across the weld. As with in-line butts the matter of angular distortion can be addressed by using differential edge preparations, welding sequence and run size. Along the middle plane of the main plate is a line of segregation or inclusions. In a highly restrained joint, for example if this detail were a 30 mm stiffening ring inside a 50 mm cylinder, precautions may have to be taken to avoid lamellar tearing under the weld. Such precautions may include using a low strength weld metal or using for the cylinder a plate with a guaranteed through thickness ductility. In flat plates, distortion is likely to arise in the continuous plate but there is little that can be done about this.

4.3.3   Corner joints

We saw in Chapter 2 that steel plate can be a far from homogeneous metal. It may contain layers of segregated constituents on a microscopic scale, inclusions of non-metallic substances on a slightly larger scale and laminations on an even larger scale. These features can cause defective joints if they are present and are not taken into account. The design of a corner joint in steel requires more careful consideration than a T joint to avoid potential defects from laminations, inclusions or lamellar tearing; the problem at the design stage is to know how far our precautions need to go because we may not know the nature of the particular steel that is to be used. To examine the position let us take a worst case scenario in which the steel is segregated and full of lamellar and laminar inclusions.

If we make a simple corner joint with the preparation on one side, as in Fig. 4.10(a), there are high residual stresses acting on the surface of plate A which can result in de-lamination if there are inclusions close to the surface or lamellar tearing. There are two means of reducing the risk of these occurrences. One is to specify clean steel, i.e. a 'Z quality' plate; this can be expensive or cause delivery problems on availability. The other is to put part or all of the edge preparation on the opposite side of the joint, as in Fig. 4.10(b). This will reduce the through thickness stress on the plate A and so the risk of defects. However with some poor quality plate there is more of a possibility of finding segregated constituents along the centre line of the surface, which may result in hydrogen cracking at the toe of the weld. It may therefore be wise to extend the bevel beyond the centre of the thickness. This precaution may also be wise in a fillet welded lap joint where the weld toe may land on a line of segregation, as in Fig. 4.10(c).

4.4   Dimensional tolerances

As with all engineering work it is necessary to define a tolerance on the dimensions of the weld preparations. This has to include not only an allowance for unavoidable small variations in the shape of the edge preparation but in the fit up between the mating parts. Such tolerances must allow for linear or angular mismatch across the joint and the welding procedure must be designed to cope with the permitted tolerances. A certain degree of mismatch caused by local variations in fit up between parts can be reduced or evened out by clamps or dogs. Note that this is not the function of a welding jig which is designed to hold in place, and to within the fit-up tolerance, parts which themselves are made to within tolerance.

Where close tolerances cannot be held, for example in the assembly on site of large sub-assemblies, the weld detail itself may be designed to accept mismatches. If a root gap cannot be held to a consistency necessary to offer assurance of a

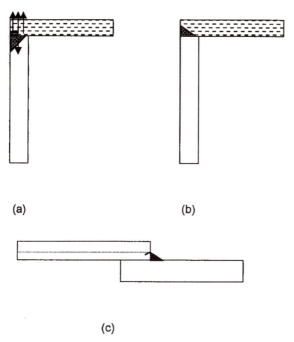

(a)   (b)

(c)

4.10 (a) Joint showing possible location of lamellar tear in plate of poor
through-thickness ductility; (b) edge preparation reduces risk of lamellar
tearing and toe position avoids segregated areas; (c) fillet weld toe
landing on centre line segregation showing possible crack.

sound weld then a device such as a temporary or permanent backing as in
Fig. 4.8 may be considered. With this approach a deliberately large root gap
is left so that full root fusion is obtained all along the joint.

To accommodate larger potential gaps between members such as in a
final closing joint, a pup piece may be cut on site to the length for the final
closure. Such inserts must be long enough to allow for two full butt welds to
be made without interfering with each other and must be the subject of a
drawing so as to provide an opportunity for authorisation and an as-built
record.

Unauthorised make-up pieces can cause problems, as in Fig. 4.11, which
shows an example in a roof truss which failed at the toe of a fillet weld
shortly after erection. The diagnosis of this was not straightforward; to the
naked eye the 'fracture surface' showed a series of lines redolent of a fatigue
fracture – not very likely in a roof truss just being erected. Closer
examination showed that these were in fact saw marks! An unauthorised
make-up piece had been inserted to make up the span, no weld preparation
had been made on the abutting ends and the weld bead had been dressed
flush. The toe of the fillet weld between the T section and the mounting plate
happened to have been placed just at the 'joint'.

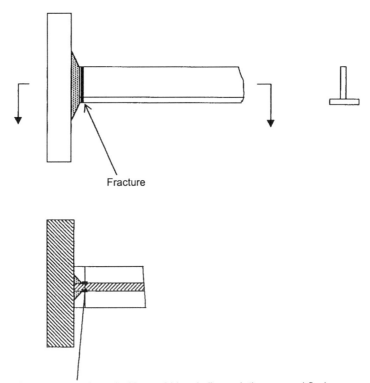

Fracture

Section square butted with a weld bead all round, then ground flush

4.11 Unauthorised make-up piece in a T section.

There have been a number of engineering catastrophes caused by yard or site 'fixes' intended to overcome mismatch or distortion and which have disastrously reduced the integrity of the structure.

In all such circumstances it is vital to ensure that the joint detail is confirmed by an engineer as being consistent with the basis of the structural design of the construction. Distortion is discussed in Chapter 11.

4.5   Access

The choice of welding preparation may have to recognise access to the joint. Access in welding means allowing the welder to see and reach the joint with the welding rod or gun whilst still being able to see the arc and manipulate it along the joint. Fig. 4.6(h) shows how the detail of the edge preparation can affect access with different types of welding equipment but the presence of

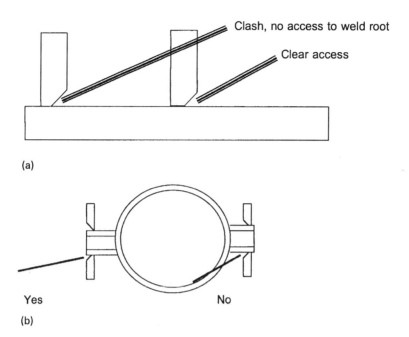

4.12 (a) Weld preparation showing need for access; (b) access to the weld root on the flange.

adjacent parts may also interfere with access; examples are shown in Fig. 4.12. With mechanised welding equipment the movement of the welding head and associated machinery must not be obstructed and the operator needs to see the arc or its position.

5

Static strength

5.1   Butt welds

Butt welds in carbon–manganese steels, made by arc welding with consumables giving weld metal matching the parent metal strength, are as strong as or stronger than the steel itself. In very high strength steels it may not be possible or feasible to produce a weld metal of matching strength and so a weld metal of lower strength than the parent metal may have to be accepted. For most purposes then, a butt weld in common structural steels does not have to be taken into account when calculating the static strength of a structure in the sense that its strength may be considered to be identical to the parent metal. However if ductility, as much as strength, is a necessary property, such as in the plastic design of a beam to column joint (Chapter 8), it is necessary to ensure that the whole joint, comprising the parent metal, its heat affected zone and the weld metal, can offer the required rotation of a section up to the specified limit state. This requires attention to the ductility of the weld metal and the ductility of the column flange material in all axes as well as to the design of the weld preparations and welding procedures so as to avoid defects such as lack of penetration and lack of fusion. These matters have a special significance in the case of earthquake resistant designs which demand extremes of ductility at the beam to column joints. The difficulties inherent in achieving such properties have been recognised and proposals have been made for the use of a beam design detail which induces the plastic hinge to occur at a position in the beam away from the welded beam to column joint.[8]

To allow high ductility to be exploited particular attention must be paid to the notch toughness of both the parent and weld metals. This may be particularly important in the case of undermatching weld metals where plastic strain may occur in the weld metal whilst the parent metal remains in an elastic state (see Chapter 2).

For welded joints in aluminium alloys, whose strength may be reduced by the heat of welding, the application standards give strengths or factors which have to be used to account for this.

5.2   Fillet welds

⊦Fillet welded joints can support loads by developing stresses which are expressed for design purposes in terms of a stress across the weld throat area. This approach postulates failure by shear across the weld throat although fractured fillet welds often exhibit different fracture positions along their length ranging from the assumed throat fracture to detachment of the weld leg from the parent plate⌡Using this simple approach for design purposes the notional stress in a fillet weld is derived from the applied load acting on the weld throat; this stress is compared with the allowable shear stress in the weld metal which, in structural steels, is often taken as half of the yield stress. The weld throat stress is calculated from the formula:

$$P/tl \hspace{4cm} [5.1]$$

where P is the applied load and t is the weld throat size and l the weld length, as in Fig. 5.1. More complex formulae have been derived from tests [9,10] and are set out in some standards.[11] These formulae are based on tests on a range of fillet weld configurations; the allowable tensile stress or resistance, depending on which approach to structural design is being used, is set as the

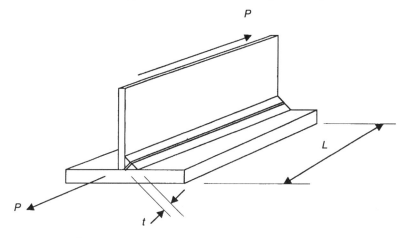

5.1   Fillet weld.

throat stress or the stress along the fusion boundary in an empirical formula. Stresses in orthogonal directions used in these formulae are labelled as in Fig. 5.2. Tests fillet welds in mild and high yield steels with nominally matching weld metal found the normal stress, $\sigma_{//}$, to have no measurable effect on the strength of the weld. This type of stress is most common in the web to flange weld in an I beam in bending. For design purposes in structural steels it was found that the three other stresses could be related to an allowable stress by a formula of the type

$$\beta\sqrt{\sigma_\perp + 3(\tau_\perp^2 + \tau_{//}^2)} \leqslant \sigma_c \qquad\qquad [5.2]$$

and $\qquad\qquad \sigma_\perp \leqslant \sigma_c \qquad\qquad\qquad$ [5.3]

where σ_c can be the allowable tensile stress or limit state stress. This is used as a basis for fillet weld design in a number of standards in which values for β are typically in the region of 0·8–0·9 depending on the strength of the parent metal.

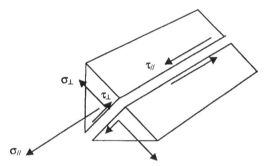

5.2   Stress notation for fillet weld.

These formulæ look very scientific and have a passing similarity to other, more formally derived equivalent stress formulae, but really they are just empirical relationships designed to fit test results. One may debate whether there is any point in pursuing a complicated calculation for a type of weld whose root fusion, and therefore throat size, cannot be confirmed, and whose results are in any case very similar to the simple method. If concern for structural integrity carries a need to calculate a stress with this apparent degree of detail surely a butt weld should be used; its integrity can be verified by conventional non destructive examination and no strength calculation is needed.

Under load, fillet welds are more ductile in shear along their length than across it but this is rarely taken into account in engineering design. In long fillet welds, as in long riveted or bolted joints, the differential strain causes the ends to take more of the initial load than the centre. No allowance appears to be made for this in commonly used design procedures. These methods all require that the weld be loaded so that there is no bending of the fillet welded joint about its longitudinal axis (Fig. 5.3). This of course could only occur with a single sided weld. Some sources say that for such single sided welds 'eccentricity effects shall be taken into consideration'. These effects are not described and their reference can cause unnecessary concern, for example in welds between hollow sections and end plates. Analyses have been published which take into account the moment induced by this

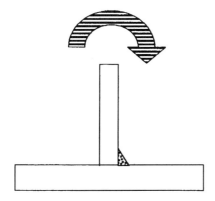

Inadmissible loading on a single sided fillet welded joint

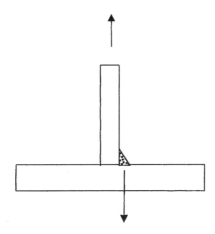

Notional eccentricity due to weld offset

5.3   Loading on a fillet welded joint.

'eccentricity' but they cannot be justified for design purposes as the results they produce are not significantly closer to the observed strengths than those of simpler methods. The design throat size, on which the strength of a fillet weld is customarily held to depend, is based on the supposition that the parts are in contact and the root is fully fused. Any lack of fit between the parts or lack of root fusion will, of course, produce a weld throat short of the nominal size. By the nature of welding such features are not normally detectable after fabrication and there are a number of ways of dealing with this matter depending on the criticality of the item. A blanket approach may be adopted in which it is assumed that all joints have a certain lack of fit/ lack of fusion of a certain amount which is then added to the nominal throat size. A more disciplined approach will be to enforce fit-up, or tolerances

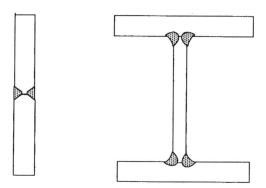

5.4 Examples of partial penetration butt welds.

thereon, and control of welding so that the notional throat is achieved in all cases. If the matter of joint strength is particularly critical then perhaps one should consider that a fillet weld is inappropriate and a butt weld, whose quality can be verified by non-destructive examination, should be used.

Partial penetration butt welds (Fig. 5.4) suffer from the same limitations in this sense as do fillet welds. Some product standards require an assumption that the actual throat is less by a certain amount than that which the edge preparation allows or the welding procedure specifies. Where an in-line joint is in compression such as a column in a building, and where there is no bending, it may be permissible for the full cross section of the member to be used even with only a partial penetration butt weld. The division of welds into fillets and partial penetration butts is an arbitrary device for the convenience of calculation, since they both represent a weld whose integrity cannot be established as readily as the butt weld. As a final point on the subject of fillet welds both the designer and fabricator must be aware that there are two conventions for stating the size of a fillet weld. One is to quote the throat thickness and the other the leg length. In the past different countries have customarily used one or other of these two conventions. This has caused problems where the fabricator and designer are in different countries using different fillet weld sizing conventions. For example a weld made to a drawing calling for a 12 mm throat fillet weld but interpreted by the fabricator as a 12 mm leg will have only an 8 mm throat. For static loading the weld will have only two-thirds of its design strength and in fatigue one third or less of the design life. Use of the ISO 2553 symbols overcomes this potential problem by requiring the drawing to state by the letter 'a' or 'z' against the weld size that it is either the throat thickness or the leg length.

6
Fatigue cracking

6.1   The mechanism

Fatigue, in the sense of the word used in engineering, is a misnomer for what is a very straightforward mechanism. What is being spoken of here is perhaps more properly called fatigue cracking; even this is an irrational term for the step by step growth of a crack under a succession of loads of a magnitude less than that which, in a single application, would not cause complete fracture or even yielding. The initiation and growth of the crack, being a function of the application of a sequence of loads, takes time during most of which the crack is invisible to the untrained eye. To early observers, the fracture from such a crack evoked ideas of a phenomenon of degradation over time which was supported in the minds of the proponents of this interpretation by the smooth fracture face redolent of a fracture in a brittle material; this led them to postulate that the material had suffered an instantaneous fracture. A phrase commonly used, unfortunately even in some quarters today, was that the material had 'become crystalline' with age; of course the metals in question had always been 'crystalline' as we saw in Chapter 2 but this was not what the proponents of this observation meant. The term 'fatigue' was adopted to reflect the loss of faculties from tiredness when viewed in terms of human experience. This nomenclature can therefore be seen to have derived from a lack of observation and analysis, circumstances more widely associated with religious beliefs where the apparently incomprehensible is given either mystical attributes which can never be fathomed or named after human conditions which are an analogy of the perceived phenomenon. What is so surprising is that these views were held by so many who should have known better so long into the twentieth century.

Probably the first published work on the subject which today is still called 'metal fatigue' or 'fatigue cracking' was in 1843 by Professor W J M Rankine in the context of some bridge girders. Then in 1871 A Wöhler, Chief Engineer of the Royal Lower Silesian Railway, published results of

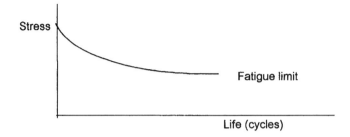

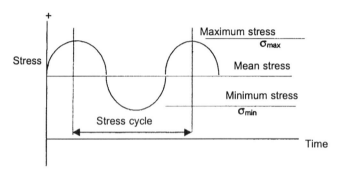

6.1 SN curve and nomenclature.

some experimental work[12] which arose from the need to solve the problem of broken wagon axles. He set up a test programme simulating the loading of the axles by testing round steel bars under rotating bending. He discovered that the time to failure of the specimens, as expressed by the number of rotations, was a function of the load; he plotted his results on a graph of load, or stress, against numbers of rotations to failure, a presentation which became known as the Wöhler. Perhaps more prosaically, it is today called an SN curve (Fig. 6.1), a graph showing the relationship of the fluctuating stress in a material against the number of repeats, or cycles, of that stress to failure. This figure also shows the nomenclature of load or stress histories.

The industry that probably did more work on the occurrence and understanding of fatigue cracking in the first half of the twentieth century was the aircraft industry. The subject was brought to the public's attention by Nevil Shute's novel *No Highway*,[13] in which an engineer, Mr Honey, predicts that the tailplane of the Reindeer type of aircraft in which he was travelling is liable to fail as a result of fatigue cracking after only a few more hours' flying. The author, whose full name was Nevil Shute Norway, was well placed to write about this subject as he had worked at the Royal Aircraft Establishment at Farnborough. He adopts acceptable artistic licence by having Mr Honey arrive at an exact calculated life in hours whereas in practice the phenomenon is set about with uncertainties deriving from load history, stress, environment, material and manufacturing detail which lead us today to quote probabilities of failure in a certain time or number of cycles. The aircraft industry was not alone in having fatigue cracking to contend with but the consequences of cracking were usually public and morbidly final compared with many other industries. In addition the efficiency of an aircraft structure, driven by the need to minimise weight, meant that stresses were proportionately higher in addition to which the early aluminium alloys were more notch sensitive than the common steels used in other products.

A vast amount of test data was acquired over the years and various presentations of that data devised in addition to the SN curve; one influence which was taken into account was the mean stress, or expressed in another way, the stress ratio, in a stress cycle. In many applications the sequence of stress is not a simple repetition of identical stress ranges and much work was done on predicting fatigue life under variable load, or stress, patterns. This led to the concept of cumulative damage in which a simple summation of the number of cycles at varying stress ranges was used as a measure of the proportion of the fatigue life used up. Various means of counting the stress cycles were invented to take account of the complexity of many load histories. In the 1950s the design philosophy was that of 'safe life' at which major structural components such as wing spars would be changed. Because of the uncertainties referred to above this safe life had to be less than the design or test life by a large factor, typically five. The design life was based on a notional flight history with load cycles based on assumed aircraft weights, flight times, gust frequencies and intensities. Naturally this was intended to provide a conservative design. It became apparent that by measuring the loads actually experienced by an individual aircraft the designers could remove one of the uncertainties and calculate the actual fatigue damage which that aircraft had sustained during a period of operation. Recognising this, the wing spars of the Vickers Viscount airliners of a number of operators were fitted with electro-mechanical strain range counters. From the records produced by these counters, the time at which a spar change was necessary would be predicted. This would frequently have been after a longer life than the nominal safe life so that a cost benefit accrued to the operator in making these measurements. Conversely an aircraft experiencing greater than usual damage could be inspected earlier than otherwise might be the case. Furthermore knowledge of the actual damage histories gave operators the opportunity of planning routes and operating procedures to minimise fatigue damage.

6.2   Welded joints

Bearing in mind that arc welding as a means of fabricating steel structures was being adopted only slowly in the 1930s it is surprising to find that fatigue tests on butt welds in structural steel plates were reported as early as 1939 by Professor Wilson at the University of Illinois in the USA. In the UK, the Institution of Civil Engineers set up a committee in 1937 to review the design practices for steel girder bridges but its work was interrupted by the Second World War and its report was finally published in 1949. The amount of interest expanded greatly world-wide after the end of the Second World War; Dr T R Gurney [14] has comprehensively reviewed the significant world literature of that and subsequent periods. The British Standard for Bridges, BS 153, then carried certain, somewhat arbitrary, requirements on reversals of stress under the passage of a live load and the civils' committee could find no logic behind their adoption. The committee examined research work around the world; German and American standards were considered to be too conservative in some respects, and unsafe in others. Arising from this review a revision of BS 153 was published in 1958 reflecting current knowledge of the fatigue behaviour of welded steel structures. This was fairly limited by comparison with today's equivalent standards. There were only two classes of welded joint, no data was given for lives greater than 2×10^6, for load cycles giving compression as the numerically maximum stress, for high tensile steels or cumulative damage. By 1959 it became apparent that the new clause could be improved in the light of new research results. More comprehensive rules were already in existence in Germany and in the USSR enabling those countries to design to higher stresses whilst using inferior steels. The Institution of Civil Engineers Committee proposed that existing data be collated and further research performed where it was seen to be necessary. This task was conducted by Gurney[15] at the British Welding Research Association (later TWI) and as a result a further revised fatigue clause was introduced into BS 153 in 1962.

In the 1960s installation and replacement of industrial plant proceeded apace in a number of industries and welded fabrications were used for the first time to replace many of the old cast, forged and riveted constructions. A rash of fatigue failures occurred as designers almost copied former shapes without understanding the effect of welded joints on the fatigue behaviour of their machines. Some of these failures had major effects on their manufacturers, some of whom were even driven out of business by the consequential losses. These problems affected ship and railway locomotive diesel engines, mine winding headgear, conveyor belt rollers, diesel railcar bogies, coal screens, earthmoving equipment, offshore drilling vessels, chemical plant mixers, overhead travelling cranes and a host of other items. But let us not be too hard on these designers, for it was only in the mid-

1950s that a Viscount aircraft crashed at Manchester because a bolt holding part of the wing flaps broke as a result of fatigue cracking. The bolt head was not squarely seated and the fluctuating bending stress set up in the shank by the aerodynamic buffeting on the flap had been sufficient to cause a crack to grow. This was only a few years after the loss in service of two Comet aircraft, the first jet powered airliner, owing to fractures in the pressure cabin from fatigue cracks. Despite the knowledge and experience in the aircraft industry, it had not been appreciated that the proof pressure testing applied to a Comet structural test fuselage prior to fatigue testing had the effect of increasing the fatigue life so the potential life of the aircraft in service was in fact shorter than had been measured in the test. This, coupled with the use of an aluminium alloy of rather poor crack resistance properties, resulted in tragic loss of life in not one but two crashes.

How the crack starts in an apparently homogeneous smooth metal is perhaps less than obvious to the casual observer but those involved in research and diagnosis of failures observed that where fatigue cracks did appear they tended to do so at sharp changes of section or at holes. This of course reflected Wöhler's observations that the fatigue life was a function of the stress. For a time it was thought that the shape of a weld cap was sufficient to create a stress concentration severe enough to start the fatigue crack but tests showed that test specimens machined out of a solid piece of steel with a cross section in the shape of a butt weld had a much longer fatigue life than the actual welds which they were intended to reproduce. It was not until Signes and colleagues[16] at the British Welding Research Association (later known as TWI) in the 1960s examined the microscopic detail of the toes of arc welds in steel that the reason became clear.

They observed that along the toe of the weld there were small irregularities rather like cracks; in reality they were tiny surface cavities filled with slag and averaging about 0·1 mm in depth (Fig. 6.2). Fracture mechanics concepts were used to predict the progression of a fatigue crack from such a weld toe and the results were found to correspond with those measured in fatigue tests. This work led to an understanding of other characteristics of welded joints under fatigue loading. It explained why, in contrast to plain unwelded specimens, there was no initiation phase in the life. Maddox[17] made growth rate measurements in plain steels of various strength and observed that for certain stress intensity ranges there was little difference in the crack growth rates. The absence of an initiation phase and the similarity of crack growth rates helped to explain why there was little difference between the fatigue behaviour of welded mild and high strength steels. This work confirmed the need to approach fatigue in welded joints very differently from that in plain steels. In plain steels there is a level of fluctuating stress below which fatigue cracking would not occur, the *fatigue limit*, which is at a stress amplitude about half of the tensile strength of the steel.

6.2 Slag intrusion at the toe of a weld (photograph by courtesy of TWI).

For welded joints there is an analogous limit, defined by the non-propagating crack size represented by the weld toe intrusions, at a much lower stress range, typically 20 N/mm^2 for all steel strengths.

In the early 1970s Gurney and Maddox[18] re-analysed the available fatigue data for welded joints and explained how some of the previously derived design data could be rationalised. Using statistical analyses supported by fracture mechanics they were able to show that some of the apparently wide scatter reported in test data was not actually random scatter but was the effect of superimposing different statistical populations each with its own, but much less, scatter. The fatigue testing programme set up in the UK in the 1970s associated with the offshore industry described in Chapter 9 paid particular attention both to the manufacturing and setting up of test specimens and the measurement of the stress in the specimen. It was found that the scatter in the results of this programme was much less than had been customarily accepted. It showed that what had been thought to have been natural scatter in previous test results had been caused in part by specimen testing techniques.

The work of Gurney and Maddox was eventually incorporated in the UK Department of Energy Guidance on the Design of Offshore Installations, which is no longer current as formal design guidance, and later in a new British Standard for bridges, BS5400 :Part 10 'Code of practice for fatigue' first issued in 1980 and which replaced BS 153 :Parts 3B and 4 :1972. However there was seen to be a need for the data to be published without being attached to or constrained in its application by being part of a product standard. Accordingly in 1993 BS 7608 'Code of practice for fatigue design and assessment of steel structures' was published which can be applicable to any product or situation.

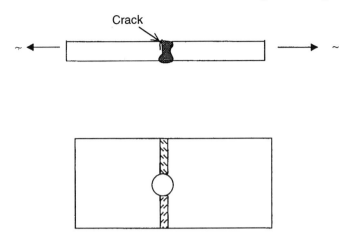

6.3 Stress concentration at a hole at a transverse butt weld.

The stress range put into the fatigue life calculation will be the nominal stress in the parts joined by the weld. The effect of the weld details on the local stress has been allowed for in these SN curves. However there will be situations where a welded joint coincides with a feature of the part which creates its own stress concentration. An example is shown here (Fig. 6.3) in which the weld comes up to the edge of a hole or where the hole was drilled through the completed weld. The elastic stress concentration in tension caused by to the hole at the weld toe will be three and so the nominal stress range in the part will have to be multiplied by three before entering the SN curve. Stress concentrations for all sorts of shapes can be found in the compendium originally compiled by Peterson.[19] The effect on the fatigue life can be quite severe, being a function of the third or even fifth power of the stress range. The shape of the joint itself can also introduce stress concentrations and most of the fatigue design data used in the world gives separate SN curves for different joint types. In this way the designer does not have to analyse the local stress distribution around the weld. It is important in using the data to define where the cracking is likely to occur. A fillet welded joint may crack at the weld toe or through the throat depending on the stress in the toe and the parent metal. The joint has to be checked for both types of crack location (Fig. 6.4). A more complicated situation where the stress at the welded joint is magnified by the configuration of the parts is in the case of nodal tubular joints.

Fatigue·cracking, as with any welded joint, will start where the local range of stress intensity is highest. The modes of loading illustrated for the tubular T joint here (Fig. 6.5) cause bending stresses as well as axial stresses in the tube wall; the highest stress at the toe of the weld will be at a position on its circumference depending on the direction of the load. This stress can

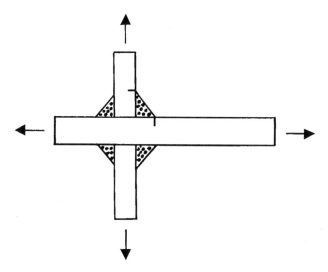

6.4   Potential positions of fatigue cracks in a fillet welded joint.

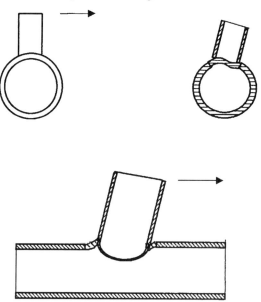

6.5   Deflections in the chord wall of a tubular joint.

be calculated using a three dimensional finite element analysis. If this is too costly and time consuming, an approximate method can be used to derive the *hot spot stress*;[20] this can then be used with the respective published SN curve to estimate the fatigue life of the joint which, as might be expected, is very similar to the SN curve for in-line butt welds. The hot spot stress is the highest stress which would be developed at a point around the toe of the weld without taking into account the stress concentrating effect of the weld profile. The hot spot stress can be found by measuring the stresses on a test specimen and extrapolating to the weld toe or by using one of a number of empirical formulae known as parametric formulae which have been published by various research workers.

The fatigue cracking behaviour of welded aluminium alloys is analogous to that of welded steels. Design SN curves have been published[21] which show that the relative fatigue behaviour of welded aluminium details is analogous to that of steels. If the applied stress ranges are compared for the same detail at the same life they will be found to be approximately proportional to the elastic moduli of the two materials.

6.3   Residual stresses

Residual stresses (Chapter 11) have the effect of placing much of the weld in an area under tensile stress which, in steels, can be considered equal to the yield stress. Any applied tensile stress will yield the material locally with the effect that regardless of the stress ratio of the applied loading the actual stress in the material adjacent to the weld will vary from yield stress tension downwards. With no effect of mean stress there is then only one SN curve for the detail, which for once makes life easier for the designer!

6.4   Thickness effect

It was known that welded joints in larger thicknesses of steel appeared to have shorter fatigue lives for the same stress history than the joints in thinner materials. This was originally put down to the greater extent of the residual stress field in a thicker section. Later, fracture mechanics considerations were to show that this behaviour could be expected for reasons relating to the crack size relative to the thickness.[15] The needs of the offshore industry for design data for thicker steels became pressing in the early 1970s. The existing fatigue rules had been based on laboratory tests, the majority of which, because of testing machine capacity were on joints in steel of around 12 mm in thickness. Many of the then new generation of offshore platforms had tubulars with wall thicknesses of 50 mm or even 75 mm. Work conducted in the UKOSRP (Chapter 9) and associated programmes had shown that this was not a straightforward matter. On detailed review, some of the tubular joint test results revealed that an apparent thickness effect could arise from the way in which an extrapolation method was used to calculate the hot spot stress at the weld toe. With this spurious effect eliminated, the true thickness effects were identified and incorporated in various current standards and codes of practice as adjustments which have to be made to be made to the standard SN curves.

6.5   Environmental effects

The rate at which a fatigue crack will grow in steels in terms of mm/cycle is affected by the local environment. An aqueous environment will increase the rate of crack growth over that in dry air or vacuum. A seawater environment is of practical interest in the design of marine and offshore structures. Seawater is a complex mixture of substances and tests have shown that its effect on fatigue crack growth in steel is not as strong as a simple saline solution. The design data for steel offshore structures makes allowances for the effect of seawater on growth rate and also the way in which it reduces the lower threshold stress. The cathodic corrosion protection systems used on marine and offshore structures can inhibit the effect of seawater on crack growth rate and in many circumstances restore it to the rate in air. Other environments can affect fatigue life and specific attention has to be paid to the material and its working environment.

6.6   Calculating the fatigue life of a welded detail

Firstly, we need to have information on the stress history acting on the detail; this may be obtained from measurements in service or from calculated stresses from the load history. In as-welded joints we do not have to worry about the mean stress; all that is needed is the stress range. This may be of a constant amplitude, which is to say that the same stress range is repeated time and time again. It does not matter whether it is repeated quickly or slowly. On the other hand each range may be different in which case it is said to be of variable amplitude.

The next stage is to find the SN curve for the particular weld detail. For design purposes various standards group weld details into categories with a similar fatigue behaviour. The type of detail has to be identified by the type of weld and also by the direction of the stress with respect to the weld; Table 6.1 shows a typical categorisation. For example the stress may be across a butt weld or along it and there will be a different SN curve for each of these situations. If the stress is at another angle the matter is more complicated but this will be ignored for this example. In some standards the weld may appear in one of two categories depending on how long it is, how close to a plate edge it is or whether the member on which the weld is made will be in bending or direct stress. It is also necessary for fillet and partial penetration welds to define at which location the crack will eventually occur – root or toe – since the lives may be different (Fig 6.4).

From the nature of the weld, the direction of the stress and the location

Table 6.1 An example of fatigue design categories of welded joints

Description of detail	Class	Explanatory comments	Examples showing crack sites
Transverse butt welds			
(a) Weld-machined flush and proved free from significant defects by NDT	C	Defect significance assessed by fracture mechanics	
(b) As-welded condition with good profile	D	Weld blends smoothly with parent material	
(c) As-welded condition other than (b)	E	Applies to welds with 'peaky' profile	
(d) Butt weld made on a backing strip without tack welds	F	The crack location is at the root of the weld	

Table 6.1 continued

Description of detail	Class	Explanatory comments	Examples showing crack sites
(e) T joint made with butt weld	F	This classification takes account of local bending	
(f) Partial penetration butt welds	W on weld throat	Check as full pen butt weld for toe cracking and on weld throat as Class W	
Load carrying fillet welds			
(a) T or cruciform joints made with fillet welds	F2 with plate stress. W with throat stress	The crack may appear at the weld toe (F2) or in the weld throat (W). At which site it will appear first will depend on the size of the weld. The joint has to be checked for both sites	

Welded attachments to a stressed member, butt or fillet weld

(a) Attachment within the width of the member, not closer than 10 mm to edge of stressed member

F

The crack will start in the member at the toe of the weld

(b) As (a) but on or within 10 mm of edge of stressed member

G

Welds in other locations, such as at tubular nodal joint

Basic SN curve T for welds. With hot spot stress

Calculate the highest local stress acting at right angles to the direction of the weld. The diagram is an example only. Depending on the axial/bending ratio the cracking may start at different places around the joint. Use SN curve for application. As an approximation class D can be used with the calculated local stress

Table 6.1 continued

Description of detail	Class	Explanatory comments	Examples showing crack sites
Fillet and partial penetration butt welds in longitudinal shear	W	Use weld throat stress	

Typical design SN curves showing welded joint classes.

of potential cracks we select an SN curve. On the vertical axis we find the stress range and reading across to the curve (actually a straight line in most log–log presentations) we read off on the horizontal axis the life in numbers of stress cycles. These SN curves are the result of statistically reducing scattered test data to a single line. This line may be the mean of the test data or it may incorporate another level of confidence. The mean line will give the life at which half the number of welds of a similar type can be expected to have cracked. This may not be thought suitable and a more conservative line, the mean minus two standard deviations, will be one in which only 2·5% of the welds will have cracked at that life. This level of confidence is commonly used as a basis for practical design.

When the consecutive stress ranges are not the same, a device called the cumulative damage rule is used. This rule was proposed in 1924 by Palmgren and restated in 1945 by Miner under whose name it is more commonly known. It is very simple and says that at any stress range S when the number of cycles of stress to failure (the fatigue life) is N then any lesser number of cycles, n, of the same stress range will use up a fraction of this fatigue life equal to n/N. This fraction is called the 'fatigue damage'; when this damage reaches one the weld has cracked, or failed. So if the stress history comprises stress ranges S_1, S_2, S_3 for n_1, n_2, n_3, cycles respectively the amount of the fatigue life used up, the damage, is

$$\frac{n_1}{N_1} + \frac{n_2}{N_2} + \frac{n_3}{N_3} \qquad\qquad [6.1]$$

which is shown in diagramatic form in Fig. 6.6. The whole life is used up when this is equal to 1. This is not an exact calculation and values in tests have ranged from less than 1 to 3 or more. Like other formulae in this book, it started life as an empirical deduction and has been shown by fracture mechanics to have some basis in material behaviour. In much of engineering, the accuracy of the stress figures as well as the stress history are uncertain and to be conservative some design authorities place a factor on this damage or on the calculated life. In some complicated stress histories

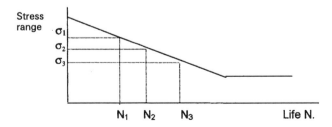

6.6 SN data for cumulative damage calculation.

it can be very difficult to decide what constitutes a stress range and there are methods of dividing up the stress fluctuations which try to conserve the intent of the cumulative damage rule; Gurney[15] reviews these methods.

Brittle fracture

7.1 Conventional approaches to design against brittle fracture

A brittle fracture in a metal is a result of crack propagation across crystallographic planes and is frequently associated with little plastic deformation. The propagation of a cleavage crack, as it is known, requires much less energy than does a ductile crack and so can occur at an applied stress much lower than that at which failure would normally be expected. In engineering materials, such a fracture usually starts from some notch such as a fatigue crack, a welding crack or lack of fusion – in other words a highly localised stress concentration. The explanation of the metallurgical mechanisms and influences surrounding brittle fracture are very complicated and the reader who wishes to know more should consult references such as Honeycombe.[22]

One of the principal reasons why the subject of brittle fracture occupies a key place in the design of steel fabrications is that the ferritic steels change their fracture behaviour with temperature, from being notch brittle at lower temperatures to being notch ductile at higher temperatures. What is more, the temperature at which this change takes place depends on the chemical composition and metallurgical structure of the steel. This is more than an academic distinction because this *transition* from brittle to ductile behaviour takes place close to the temperature at which many steel fabrications operate. The phenomenon is associated particularly with welded fabrications because the energy required to propagate a brittle fracture is low which means that the stress required to start the crack can be supplied just by the residual stresses from welding without the necessity of an externally applied stress. Welds may supply a stress concentration in the form of a crack. Furthermore welding can damage the fracture toughness of the steel, and in the past some weld metals themselves had very poor fracture toughness. A brittle fracture can be driven by the strain energy locked up in the metal and may not need an external load or force to start it. Brittle fracture is a fast moving unstable fracture which has been known to sever complete sections of welded bridges, ships, pressure vessels and pipelines. The speed of the progression of the crack front has been calculated as about half the speed of sound in the steel. In some cases the crack has been arrested by the exhaustion of the strain energy or by its running into a region of high fracture toughness. The basis of the approach to design and fabrication to prevent brittle fracture occurring then lies in appropriate material selection and welding procedure development.

In a limited number of applications steps are taken in design to introduce devices which will arrest a running crack. For example in pipelines and other vessels the longitudinal welds in adjacent pipe lengths are offset to avoid presenting a continuous path of similar properties along which a fracture could run. As an alternative to this, a ring of thicker material or higher toughness material may be inserted at intervals which locally reduces the stress sufficiently to arrest a crack. It has to be recognised that it requires a material of much higher fracture toughness to stop a crack than would have been necessary to have prevented it starting in the first place.

We should recognise that the consequences of service can also lead to circumstances where a brittle fracture may occur in a fabrication which was initially sound. For example fatigue or corrosion cracks may grow to a critical size during the life of the fabrication; irradiation in nuclear plant can reduce the fracture toughness of steels. Materials other than ferritic steels need to have defined fracture toughness but they do not exhibit a significant change of that property with temperature and so the question of material selection has one less dimension. We shall see later on in this chapter how the steel can be tested to classify its suitability for use in any particular circumstance but first we need to consider the factors that have a bearing on the requirements for fracture toughness.

For any given quality of fabrication these are:

- thickness
- applied stress
- fracture toughness.

The criterion of applied stress referred to here is not a question of small differences in calculated stress in a member but whether or not there are large areas of high stress concentration and constraint. Examples of these areas are the nodes in tubular joints where there are large local bending stresses, caused by incompatibility of deformations, and the stress concentrations inherent in openings, nozzles and branches in pipelines, pipework and pressure vessels. Greater thickness is a feature which engenders tri-axial stress systems which favour plane strain conditions. In addition thicker material will contain more widely spread residual stress systems than thinner material. For any combination of thickness and stress we can then choose the level of parent metal fracture toughness which research and experience has shown to be appropriate. Perhaps it is not unexpected that the appropriate choice will be set down in a standard specification for the product or application which we have in mind and which itself will refer to a range of steel specifications in another standard. The application will also perhaps introduce as a basis of selection other criteria which have not been mentioned so far such as that of risk, represented by the hazards, their consequences and the likelihood of their occurrence.

7.2   Fracture toughness testing and specification

Incidents of brittle fracture in riveted structures were reported in the late nineteenth century and of welded structures in the 1930s but no coherent approach to investigating the reasons emerged. Eventually it was the fracture of the hulls of more than one fifth of the nearly five thousand 'Liberty Ships' built in the Second World War which led to work on categorising welded steels by their propensity to brittle fracture.[23, 24] The Liberty Ship was a type of merchant ship, virtually mass produced by welding in the USA in response to the need to keep the UK and the USSR supplied with fuel, arms, food and other necessities in the face of the German attempt to blockade the North Atlantic and other sea routes using submarines. Investigations in the USA and the UK concluded that incidents of brittle fracture in these ships were more likely to have occurred where the Charpy test energy of the steel was less than 15 ft lb (20 J). Even today most structural steel specifications use this measure of fracture toughness, even if not the same numerical value. In the Charpy test a notched bar of the steel is struck by a pendulum (Fig. 7.1). The energy absorbed by the bending and

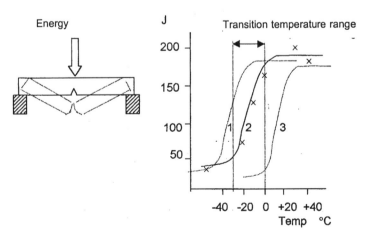

7.1 Charpy test specimen and typical results.

fracturing of the bar is a measure of the fracture toughness of the steel. These tests are done on a number of samples at different temperatures and the energy absorbed is found to vary with the temperature. The change of energy occurs over a range of temperature called the *transition temperature range*. The energy measured is not a fundamental measurement which can be mathematically related to quantities such as stress intensity although certain empirical relationships have been derived. However as a result of experience, certain minimum values of Charpy test energy have been found which give freedom from brittle fracture in conventionally fabricated constructions. The full line, 2, is the curve given by the set of results marked X. The higher results to the right are on what is called the *upper shelf* although the minimum values required by many specifications will often be found in the transition range but above the *lower shelf* figures. The steelmaker can produce steels with different levels of fracture toughness and different transition temperature ranges as in lines 1 and 3. Within the carbon–manganese steels this is achieved by a combination of metallurgy, mechanical working and heat treatments. Generally the finer the grain size of the steel and the fewer the non-metallic inclusions the higher will be the fracture toughness. This property in the parent material determines the lowest temperature at which a fabrication can be used, provided that it is not overridden by the weld and heat affected zone properties. The minimum temperature at which it is practical to use carbon–manganese steel fabrications is around –40°C. The alloy steels containing around 9% nickel are suitable down to around –190°C. Below that temperature austenitic steels or aluminium alloys can be used. Although they exhibit no sharp transition temperature effect their fracture toughness still has to be controlled.

The temperature at which the minimum Charpy energy is specified is not necessarily the minimum temperature at which the fabrication can be used safely. The Charpy test specimen is of a standard size, 55 × 10 × 10 mm, regardless of the thickness of the steel from which it is taken. (There are sub-standard sizes for materials thinner than 10 mm.) The effects of thickness which we have spoken about mean that as the thickness goes up we have to use a steel with the minimum required Charpy energy at lower test temperatures. For example, a typical offshore platform specification might require a certain minimum energy level at certain Charpy test temperatures, depending on the minimum service temperatures, for as-welded, i.e. not post weld heat treated, fabrications in regions of high stress. These regions would normally be the nodal joints in tubular structures. For other parts of the structure less demanding properties might be required. These temperatures apply to carbon–manganese steels of all strengths and a different Charpy energy is required of each grade of steel. Typically the minimum energy required is equivalent numerically in Joules to one tenth of the highest minimum yield strength of that grade of steel in N/mm^2. This is necessary because the energy required to bend a Charpy test specimen prior to fracture in a higher yield strength steel will be greater than that required for a lower strength steel.

Other products such as buildings and bridges have their own require-ments which are usually less demanding than those for offshore construc-tion; they recognise the stress concentrations, service conditions, the consequences of failure and the customary levels of control in the respective industries. These requirements are expressed in various ways. In some products the required Charpy test temperature for the steel is related to a range of thicknesses. In others the thickness of the steel requires a certain steel grade without direct reference to a Charpy value or some other measure of notch toughness.

In summary the Charpy test has a number of limitations. As well as being conducted on a thickness of material not necessarily representative of the structure in question, the test measures both the energy absorbed in bending and then fracturing the specimen; further, it is carried out at a high loading rate unrelated to most service conditions. These features mean that it is not a basic measure of the ability of a material to survive and the results cannot be interpreted in a quantitative way. Nonetheless it has the benefit of using easily made and repeatable specimens and the test itself is simple and quick. It is therefore a valuable quality control tool.

7.3   Fracture mechanics and other tests

Where a more discriminating test than the Charpy test or one giving results which can be applied to the assessment of defects is required, a fracture mechanics test can be used. Such tests can use a specimen from the full thickness of the material under study and with a crack starting notch which is more representative of actual weld defects than the rather blunt notch of the Charpy specimen. The state of stress around the tip of a sharp crack can be described by a quantity known as the stress intensity, K_1. In a fully elastic material this quantity may reach a critical value at which fracture occurs, K_{1C}. We can measure this by carrying out a fracture mechanics test which entails bending a cracked specimen and measuring the load at which fracture occurs. By calculating the stress at the crack tip at fracture the value of K_{1C} can be calculated. This can be used to make an assessment of the significance and acceptability of weld defects or fatigue cracks if the stresses are elastic.

However we have seen that structural steels are far from being elastic when they reach yield point. The stress distribution around a crack or other weld defect is extremely complicated especially when plasticity comes into play. In the welded joint the residual stresses are a significant part of the overall stress pattern. To investigate this situation a test on an actual weld in a realistic thickness of plate was needed. This led in the 1960s to the development of the Wells wide plate test and the crack opening displacement (COD) test at the British Welding Research Association (BWRA, later to become TWI).[23] Welds with artificial defects were made in a test plate to give a specimen about 1 m square which itself was welded into cast end pieces through which loads from hydraulic capsules were applied, eventually up to 4 000 tonnes. The COD test involved taking a sample of the similar type of weld as a coupon and cutting into it a notch. This specimen was then put in bending whilst the opening of the notch was measured until fracture occurred. The test was further developed with more refined measuring techniques and the extension of the notch by fatigue cracking (Fig. 7.2). This gave the finest possible notch and one which could be produced consistently. Even at low applied stresses the crack tip actually stretches plastically and this can be measured as the crack tip opening displacement, δ (CTOD). The value of this as measured at fracture is used in assessment of the significance of cracks or other features, particularly in welded joints. In preparing the specimen the notch is first sawn and then grown by fatigue cracking to produce the finest possible and most consistent crack tip. The opening of the notch is measured by an electrical displacement gauge and the actual tip opening is calculated on the basis of the crack and bar geometry. The use of a fatigue cracked notch not only ensures that the finest crack is produced but it can be placed within the cross section of a welded joint so as to sample quite narrow regions of a particular microstructure in the weld metal or the heat affected zone. In Fig. 7.3 a 'K' preparation has been used to give a heat affected zone straight across the section so that the fracture will always be within the same microstructure as

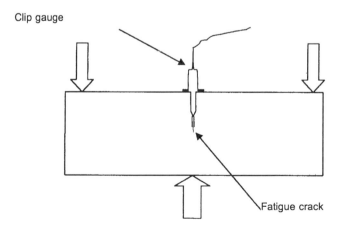

7.2 Specimen set up for crack tip opening displacement test.

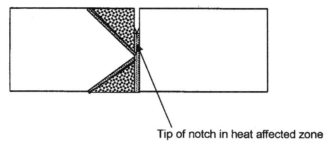

Tip of notch in heat affected zone

7.3 CTOD test specimen for a butt weld showing the tip of the notch in a weld selected.

7.4 A brittle fracture (photograph by courtesy of TWI).

it moves into the specimen. Fig. 7.4 shows a brittle fracture surface with the typical chevron pattern 'pointing' to the originating crack.

8.1   Structural forms

8.1.1   Steel frames

Steel building frames range from simple single-storey buildings to vast multi-storey skyscrapers. They have bolted, riveted or welded joints and attachments. One of the earliest of the large iron building frames was the Crystal Palace built in London for the Great Exhibition of 1851.[25] This was designed by Joseph Paxton, who was not an engineer. He was initially a gardener, becoming head gardener at Chatsworth, the seat of the Duke of Devonshire whose grounds were laid out by Capability Brown. Paxton eventually became a director of the Midland Railway. The Building Committee of the Great Exhibition included engineers of the eminence of Brunel and Stephenson. They accepted Paxton's design in preference to the Committee's own design which, like most committee outputs, was the lowest of common denominators. Since Paxton had no engineering, knowledge the detail design and calculations for his concept were performed by the contractors, Fox Henderson & Co of Smethwick. Fox was later to found the firm of consulting engineers that became Freeman Fox and Partners in the twentieth century and that was responsible for some of the large bridges in the world today. Sample elements of the Crystal Palace structure were tested and survived four times the design load before fracture. The structure relied for its lateral stability entirely on the rigid connection between vertical iron columns and horizontal beams. In this it differed from all previous iron constructions in which this portal bracing had been achieved either by arched girders or spandrel brackets. In this manner it reflected the basis of future beam and column structural design which has been used for most building frames since.

8.1.2 Box sections

In this context we are speaking of built-up rectangular box sections and not rolled hollow sections, which are covered in the next section. Boxes are a very efficient section for long bridges. They are relatively easy to build and paint and the interiors can be used for access for inspection and repair as well as for carrying services. Being in effect large and relatively thin plate structures distortion has to be controlled and particular attention has to be paid in the design to structural stability to prevent premature buckling. This instability led to a disaster when a diaphragm in one of the lengths of the box section of the new bridge for a motorway at Milford Haven near Bristol collapsed as it was being rolled out over a support. This incident followed closely on the collapse of another box girder bridge during erection, the Yarra Bridge in Melbourne, Australia. This comprised two parallel boxes curved in plan. In attempting to rectify a mismatch in elevation some flange bolts were taken out of one box but this allowed the top plate to buckle and the structure collapsed and fell. The causes of the two collapses were different but both being box girders designed by UK consulting engineers, an enquiry, the Merrison Enquiry, was set up in the UK to examine the whole matter of welded box girder design. An outcome of this enquiry was that the design practices were changed to take account of the effect on stability of residual stresses and dimensional tolerances in large thin panels.

Residual stresses and distortion are two sides of the same coin and both can affect the ability of a plate to carry a compressive load.[26] A plate in compression will support a load up to a point where it begins to buckle. The stress at which buckling starts in a perfectly flat plate is a function of its thickness, the width and length between members bounding the plate and the proportions of the boundary members themselves.[27] When the plate buckles it can no longer support the load which is then taken by the boundary members which themselves may be unable to support the load. Fig 8.1 shows a simple panel in compression. The theoretical buckling stress is given by an equation of the form:

$$\sigma_b = K \frac{E}{1 - v^2} \left(\frac{t}{b}\right)^2. \tag{8.1}$$

K has values depending on the ratio of the length of the sides and the fixity of the edges. If the plate is welded onto the boundary members there will be tensile residual stresses along its edges which will be balanced by compressive stress in the centre of the plate. The result will be that the applied load required to cause the plate to buckle will be less than for a plate without residual stresses. Further, if welding has caused the plate to distort out of its plane it will buckle earlier than would a perfectly flat plate. These effects are taken into account in setting design stresses for welded plate and

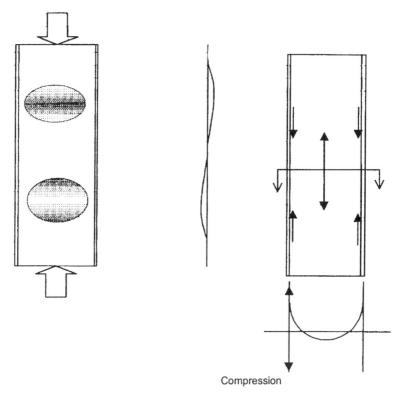

Compression

Residual stress disribution

8.1 Plate buckling.

box structures. Clearly it is important to structural performance that the residual stresses and distortion are kept as small as possible by careful design of the structure, the welding procedures and the planning of welding sequences.

8.1.3 Tubular members

8.1.3.1 Early examples

In a surprisingly short time after iron and steel began to be used as a structural engineering material, tubes were adopted as a structural form in some very large structures. Amongst the earliest examples of large scale tubular steel structures were three railway bridges in the British Isles. In 1848, Robert Stephenson built the Britannia Bridge to carry the railway across the Menai Straits between North Wales and the island of Anglesey. This bridge is a rectangular box section, another form of tube, or hollow section as we might call it today. The Tamar Bridge, opened in 1859, was the work of Isambard Kingdom Brunel, well known for his other engineering works, and carries the main line from London to the West of England across the River Tamar at Saltash near the south west coast of England; it is a two span bridge in which each span has a curved oval section tube as a top chord. The Forth Bridge, Fig. 8.2, the work of Sir John Fowler and Sir Benjamin Baker, carries the two tracks of the main East Coast railway line between London and the north east of Scotland across the Firth of Forth in Scotland; its construction was started in 1882 and it was opened in 1890. It sports tubular members on a grand scale and amongst other things it is notable that as a contribution to structural integrity the rivet holes were reamed. These structures do not strictly come within the scope of this book because they are not welded but they do illustrate that the tube was a structural form whose properties were appreciated by some of the greatest engineers of the past.

On a much smaller scale than these grand bridges of the nineteenth century, steel tube began to be used for bicycle and motorcycle frames in the nineteenth century and for many years the tubes were joined mainly by brazed socket joints although welding has since taken over on motor cycle frames and some cycle frames. The first welded production motor cycle frames were made with MAG welding in the 1960s and suffered early fatigue cracking. The designers had not realised how good was the fatigue performance of the old brazed socket joint which has the other benefits of being self jigging, easy to paint and easy to clean in use because of the

8.2   The Forth railway bridge.

smoothness of the brazed socket. The low temperature of the brazing process also allowed alloy steels to be used without loss of their strength.

Tube was used for major components of many of the early aeroplane fuselage and wing structures, even until the 1940s, in airframe components such as the fuselage of the Hawker Hurricane, first flown in 1935, and the wing spars of the Vickers Wellington which first flew in 1936. The Wellington spar was of aluminium alloy tube which at spanwise wing joints was connected by serrated plates clamped in place by transverse bolts, a detail which today would raise concerns about fatigue performance. After that period the only major items in aircraft made of tube have been engine mountings and some light aircraft fuselages and helicopter tail booms mainly constructed of welded steel tubes. Such structures were originally made by gas welding the joints which suited the small sizes of tube and gave smooth joints. They perhaps have a better fatigue life than the same joints made with metal-arc welding and which in later years have been reproduced with TIG welding.

8.1.3.2   Tubulars in buildings, offshore platforms and other structures

The tube, or hollow section, has been used by man from time immemorial as supplied by nature in the form of bamboo. Even today in industrialised South East Asian countries bamboo is used for quite large scaffoldings around buildings; the joints are made with lashings made of plastics in place of traditional vines or grasses. Since the middle of the twentieth century steel tubes have been used extensively for structural purposes not only as circular hollow sections but increasingly as square and rectangular hollow sections which have found favour in buildings, small bridges and other architectural applications where their properties and appearance gave them advantages over the traditional rolled steel joist, I and H sections. Extensive research into the properties of joints in these hollow sections has been funded by the steelmakers as part of their marketing strategy. This has led to a detailed understanding of the performance of welded joints in hollow sections, and the development of optimum configurations of the joints for various load combinations. Most steel for hollow sections used in buildings is carbon–manganese steel, although a rather unusual building in Cannon Street in London has an exposed tubular lattice made of ferritic stainless steel tube whose members are filled with water for fire resistance.

Oil drilling and production installations have been constructed since the early part of the twentieth century. As exploration and production moved from dry land to swamp to lake and then to the open sea, the drilling rig and then the production equipment had to be supported above the water on the type of platform which has become so common today and which was initially developed for use in the Persian Gulf, as the Arabian Gulf was then called, the Gulf of Mexico and South East Asia. These platforms are constructed mainly of steel tubes with welded joints and this subject is expanded on in Chapter 9. A whole branch of structural engineering practice grew up around them, eventually being embodied in standards and codes of practice such as RP 2A published by the American Petroleum Institute. The necessary diameters and wall thicknesses of the tubes at the point where they met each other, nodal joints as they became to be known, were related to the loads through simple and empirical formulae such as *punching shear*[20] later to be refined by the hot spot stress concept.

8.1.3.3 Designing tubular joints

From an early stage in their training structural engineers are taught to avoid designing into their structures eccentricities and out-of-plane loads because they set up local bending (secondary) stresses in addition to the primary stresses. Primary stresses are those stresses calculated by the conventional global methods of structural analysis but calculating secondary stresses requires more detailed methods such as those using finite elements. The effect of secondary stresses can lead to local instability or plastic collapse under loads lower than the design loads or, in the case of fluctuating loading, a shortened fatigue life. These secondary stresses are customarily avoided by the simple expedient of designing members to transfer loads in line or by introducing back-up members across plates. Examples can be seen in the design of bridge girders over the supports and the girders of topside modules of the big offshore platforms where there are 'stiffeners' or back-up members in the plate girders where the transverse loads are reacted (Fig. 8.3). Historically this concept was not adopted on most tubular nodal joints. In these, a joint was made where two or more tubular members meet by standing the ends of the braces on the surface of the chord. This places the chord wall in bending which will be seen to contradict the structural engineer's training and really ought to be seen as downright bad practice (Fig. 6.5).

How was it then that the designers of the tubular structures made nodal

8.3 Detail of heavy girder constriction showing back-up members.

joints between tubes by placing the end of one tube against the unsupported wall of the other so developing local bending stresses? Why do these designers of tubular structures not follow the good practice well established for decades if not centuries? The answer may lie in the old human qualities of conservatism and lack of vision, or, in the vernacular, they couldn't see the wood for the trees. So let us look at the current design practices design for tubular joints whether for a building, an offshore structure or a road vehicle. The first step is to decide what shape and size of tube is to be used. This cannot be done for each member in isolation. A feature of tubular structure design is that the joints tend to control the relative member sizes. In general we start with the main members whether we call them columns, legs or chords. Their size will depend on the load they are expected to carry either statically or as a fluctuating load. Local buckling will decide the proportions of the cross section which may or may not have to be stiffened. Overall buckling will influence the spacing of bracing members. The size of these braces may well depend on the joint which has to be made between them and the column or chord. For architectural uses, the selection of relative member sizes at the joints may be based on appearance rather than their structural performance, which of course still has to be adequate.

8.1.3.4 How tubular joints work

We can start with a simple T joint between circular tubes consisting of a chord onto which is fixed a brace at right angles. It is a simple symmetrical joint which will help to explain how tubular joints work in general. When the brace is loaded axially, i.e. along its length, the force is resisted by the chord. Fig. 8.4 shows how this transfer of load occurs. When the two tubes are of equal size, most of the load transfer takes place at the flanks where the joint stiffness is highest. When the brace is very much smaller than the chord it tries to punch through the chord and its load is resisted by the shearing force through the chord wall which distorts under the load under the local bending effect. The distorted shape of the chord wall is controlled by its being attached to the brace, and the loaded member is itself acting as a stiffener, so there is a very complex pattern of stresses set up both in the chord wall and in the end of the brace. When a brace the same size as the chord is loaded laterally, in the plane of the joint, the bending load at the chord is resisted by shear at the flanks and also by shear in the chord wall elsewhere; if the load is out of the plane of the joint the load is resisted more at the flanks than elsewhere. If the brace is much smaller than the chord, the chord wall is put under higher local bending and shear from either the in-plane or out-of-plane load. What is particularly significant is that it is at these points of high stress that the welds are placed.

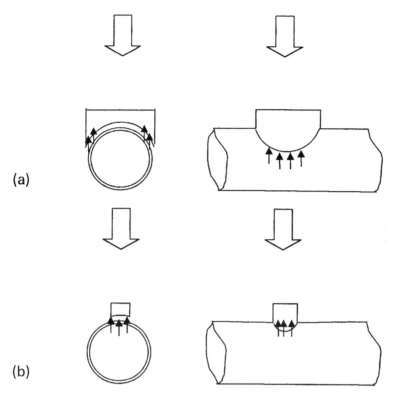

8.4 (a,b) Brace and chord of equal diameter, load reacted mainly on chord flanks; brace smaller than chord, load reacted on chord face.

8.1.3.5 *More detailed information*

There is a small number of authoritative works reviewing the knowledge of tubular joint behaviour and design in detail and the reader who wishes to read more will find the following works of great value.

In 1982 Professor Jaap Wardenier of Delft University in the Netherlands published a comprehensive work[28] on the design and performance of hollow sections in steel presenting the outcome of research across the world on the static strength and fatigue characteristics of joints in both circular and rectangular hollow sections and in joints between circular and rectangular sections and between circular braces and open rolled section chords. Some ten years later Dr Peter Marshall of the Shell Oil Company published a commentary on welded tubular connection design.[21] This work was written to explain the basis of tubular joint design as expressed in the American Welding Society's Structural Welding Code D1.1 to those engineers who had not been involved in the development and application of the experience in designing tubular structures for the offshore industry. For those engineers who just want to know what to do when designing a structure with hollow sections for conventional buildings or similar purposes the British Steel publication *SHS Welding*[29] gives the necessary details and allowable stresses based on BS 5950 'Structural use of steelwork in building'. Taken together these three works could be said to encompass most of the background to the design of welded tubular joints and it would be superfluous to reproduce the detail here.

8.2   Design philosophies

8.2.1   Elastic method of design

Traditional structural steel designs were based on the idea that if the calculated stress in any part of the structure did not exceed an *allowable stress* then it would safely support the load it was designed to carry. This allowable stress, or working stress as it is sometimes called, in both tension and compression was set as a fraction of the yield stress or tensile strength. For pressure vessels and some structures this allowable, or design, stress was once set at a quarter of the ultimate tensile strength but later this was changed to two-thirds of the yield stress. This approach is called the elastic method of design because under the design load nowhere in the structure is the nominal stress intended to exceed the yield stress; but clearly there will be stress concentrations at bolt and rivet holes and other openings where the stress may be up to yield stress. Whilst such concentrations are accepted in framed structures such as buildings and cranes, pressure vessel design

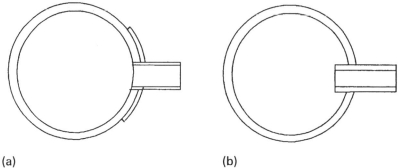

(a)   (b)

8.5 (a,b) Reinforcing of the shell by a doubler plate around a nozzle; reinforcement of the shell by the nozzle itself.

practice is to introduce reinforcing at openings to avoid generating large areas of yielded material; this reinforcing may be in the form of a plate or as the branch or nozzle for which the hole exists (Fig. 8.5). In bridges, cranes and other types of structure subject to fluctuating loads, bolted joints are designed so that the stress concentrations are taken into account in the working stresses to avoid premature fatigue cracking. For members in compression buckling is avoided by a reduction in the allowable stress depending on the length and cross-section. For some transient loading conditions such as wind gust loads on buildings the maximum stress can exceed the normal allowable stress by perhaps 25% on the basis that the structure will not have time to react dynamically in the brief period for which the load exists.

8.2.2 Plastic theory of design

Although being a simple method of design the elastic method produces rather inefficient steel frame structures in terms of the weight of steel used to support a load. The size of a member is based on the maximum moment anywhere in it; for a simply supported beam with a distributed or point load this is at only one point on the beam. The result is that the remainder of the beam is increasingly over-designed towards its ends. By making the end joints rigid, the maximum moment is reduced and moments are introduced at the ends; a smaller beam section can then be used and more effectively since more of its length is working nearer its design strength (Fig. 8.6(a)). Beyond this, rigid joints offer a further opportunity in steel which was to be exploited by what was to become known as the plastic method of design. This was developed in the 1930s by J F Baker (later to become Lord Baker of Windrush) and colleagues at Bristol University under the aegis of the Steel Structures Research Committee. This method was based on the observation that a rigidly jointed structure would not collapse until sufficient members had plastically deformed in such a way as to form a mechanism (Fig. 8.6(b)). This occurred when at the points of maximum moment the whole section would yield and act as a hinge, a *plastic hinge* as it was called. The corollary of this was that in deforming plastically, the steel would absorb energy.

In the event, the first practical application of the plastic method of design was not to be as originally envisaged, in building frames, but in a type of domestic air raid shelter, the Morrison Shelter, introduced in Britain in the Second World War. Until that time the shelters commonly used by individual families as some protection against German air raids all over the British Isles were Anderson Shelters. These were dug-outs in gardens reinforced with an arch of corrugated steel sheeting which was covered with earth. In the inner city areas, where there were flats and office buildings with

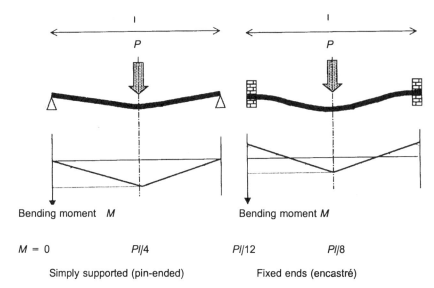

Bending moment *M*   Bending moment *M*

M = 0   *Pl*/4   *Pl*/12   *Pl*/8

Simply supported (pin-ended)   Fixed ends (encastré)

(a)

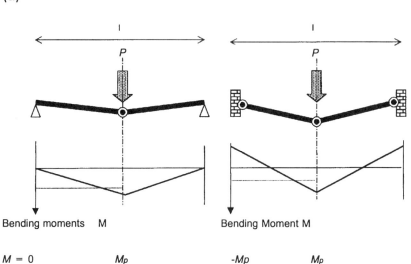

Bending moments   M   Bending Moment M

M = 0   *M*p   -*M*p   *M*p

Simply supported (pin-ended)   Fixed ended (encastré)
beam collapses when centre   beam collapses when
moment = *M*p and plastic   end moments = *M*p
hinge forms.   and plastic hinges form
                                      at the ends and centre.

(b)

8.6(a,b) Elastic bending moment in a beam under a point load;
plastic bending moment in a beam under a point load.

no convenient gardens and in schools, both in town and country, where there were too many children to use a dug-out, brick-built communal shelters were constructed on the streets and playgrounds. In London these supplemented existing underground spaces such as underground railway stations which were used as shelters. The effectiveness of such arrangements in protecting life relied on forewarning people of raids in the daytime so that they could take cover in the nearest shelter and on their sleeping in them at night. Even so the tragic facts are that during the first two years of the war, up to the end of 1941, the 190 000 bombs, both explosive and incendiary, dropped by German aircraft on Great Britain killed some 44 000 civilians, including 5500 children; they seriously injured 50 000 people, 4000 of them children.[30] Later in the war manned German bombers were replaced by the V1 or Flying Bomb, also known more informally as the Doodlebug and which, in the light of future weapons, has since been called the first cruise missile. These were sent over the south east of England in 1944 and 1945 in thousands mostly aimed at London but in practice falling over a large area of southern and eastern England. Their small size, speed, number and unpredictability of the site of their eventual fall to earth made any form of useful advance warning impracticable. Totally unpredictable was the later V2 ballistic missile plunging to earth at supersonic speed. Each of these types of weapon carried a 1000 kg high explosive warhead whose effect, as with bombs dropped from aircraft, was not only to kill and maim people and demolish the buildings near where it fell but to radiate a blast wave which would typically suck a wall out of a house so removing the support to the floor joists leaving the floor to collapse as a slab.

Having to sleep in underground stations or other shelters was not really satisfactory for long periods so the Morrison Shelter was developed, named after Herbert Morrison the then Home Secretary. It was colloquially called a table shelter. It was actually installed inside houses, giving people a shelter whether or not they had a garden and so enabled them to remain in their homes, albeit still at some considerable risk of death or injury from bombs. This shelter protected the occupants of a house from flying and falling debris but more specifically from the collapse of the upper floors. Shown in Fig. 8.7, it comprised steel portal frames in two planes covered on the top by a steel sheet and on the sides and base with wire mesh; ordinarily it served not only as a table but as a bedstead. The components, a steel sheet and a number of pieces of rolled steel angle with bolt holes, steel mesh and bolts could be assembled quickly by unskilled labour, if necessary the recipient family themselves. The family could take refuge in the shelter in the event of an air raid warning and could place a mattress in it on which to sleep at night. Had the frame been designed on the conventional allowable stress basis it

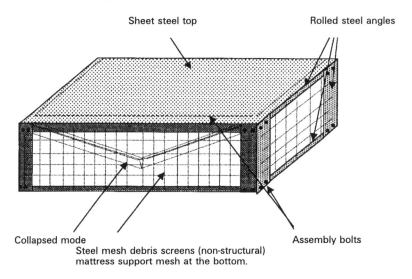

Sheet steel top

Rolled steel angles

Collapsed mode

Assembly bolts

Steel mesh debris screens (non-structural)
mattress support mesh at the bottom.

Approximate dimensions 2m × 1.5m × 0.75m.

8.7 Principal features of a Morrison Shelter.

would have been far too heavy to have been supported by the timber floor of a house. Since the shelter was intended to survive only a single event, the member sizes and the corner joints were designed so that the frame would protect the occupants by partially collapsing in a controlled plastic manner so absorbing the energy of the descending floor rather than by offering rigid resistance. This allowed the members to be lighter than the conventional design practice would allow. A shelter was delivered to each household as a kit of simple steel parts with pre-drilled holes for bolts; it was assembled in a room on the ground floor or in the cellar of a house.

After the war, the application of welding for steel building frames offered a much greater opportunity for the exploitation of the plastic design method. One of the requirements of such a structure was that the joints should be able to develop the full plastic moment of the beams or columns, a characteristic which welding was particularly able to produce. Baker had by then become Professor of Engineering at the University of Cambridge and with his colleagues developed this design method[31, 32] which was first used for the steel frame of a school at Hunstanton in Norfolk. The second building in which it was used was the Fatigue Laboratory at The Welding Institute. The plastic theory is not applicable in all circumstances, for example where deflection or fatigue life is a constraint, and a useful commentary will be found in the *Steel Designers' Manual*[33] and in the book by Davies and Brown.[34]

8.3   Limit state design

The plastic design method is an example of what we now call *limit state design*. This approach to design is based on the definition of a condition, or state, of the structure beyond which it will not be allowed to go. If this state is for the normal service in which the structure is neither to deflect more than a certain amount nor to show any permanent deformation of the members then it may be called the serviceability limit state. If the state is to be defined in terms of partial or complete collapse of a frame, for example by yielding or buckling, it can be called the ultimate limit state. The method can be applied with other criteria such as fatigue cracking or structural oscillation or resonance. It is very different from the elastic design method in which the stress is calculated not to exceed some arbitrary value which may not have a rational relation to the actual load bearing capability of the structure.

The limit state design procedures can place factors on the material properties to allow for natural variations in those properties. Factors can be put on the loads to allow for the probability of each type and size of load occurring. It thereby can be a much more discriminating design process for some types of structure and has the potential for producing more efficient and optimised designs.

9.1   The needs of deepwater structures

The development of welding design philosophies for deepwater offshore structures took place over a very short time and so deserves a chapter to itself providing a compact scenario of how such philosophies can evolve. The structural hollow sections used in buildings, cranes and so on are relatively small compared with those which came to be used in the construction of deepwater offshore platforms in the North Sea in the 1970s but there is a commonality of approach which has benefited both scales of structure in providing a rational design methodology throughout tubular joints.

The design of welded nodal tubular joints, particularly for offshore structures, which was already well established for shallow waters, became almost a specialist sub-discipline of engineering in the early 1970s. In Europe this situation may be said to have started in the mid-1960s when the North Sea exploration off the coasts of the Netherlands and the UK had found gas; but as exploration moved north opportunities for oil production were revealed. Both the water depth and the extremes of weather increase to the north. The much greater water depth than was customary in the southern North Sea, the Gulf of Mexico, the Persian Gulf and the Far East from which most offshore oil and gas had so far been extracted would require platforms of a much larger size. Their size was not only a response to their environment directly; their isolation and so the distance over which the product had to be piped gave the need to conduct some preliminary processing of the product which required on-board plant. The isolation also required a fairly large resident crew with appropriate quarters and stores for long stays, and a helideck capable of accepting the largest helicopters in use. After a journey of one and a half hours, they required refuelling from a store of aviation fuel on board the platform. This resulted in a platform which was more a multi-storey hotel and office block cum process plant than the traditional spindly shallow water platform with one or two operating cabins and a few bunks only half an hour's helicopter flight from land.

The drive for the development of new platform designs was the urge to produce oil as soon as possible after the discovery of viable quantities in the North Sea. In addition to their size other considerations were the more continuous occurrence of larger wave sizes and lower temperatures than in previous locations. Lloyd's Register of Shipping had been reviewing the position with oil companies such as Shell and BP and jointly concluded that there was insufficient valid information on which to pass judgement on the long term integrity of steel structures which were being designed for the northern North Sea oil fields. They felt that it would be unwise to extrapolate the design rules used for the joints in the smaller platforms and apply them to the larger ones. Two areas in particular in the structural field were deemed to require attention, namely fatigue and fracture. The importance of such matters had been tragically demonstrated in a disaster which had occurred in the early stages of North Sea oil and gas development. In December 1965, 13 men died when the self elevating barge *Sea Gem* collapsed in the North Sea. This catastrophe was to have a large and permanent influence on the approach to material specification and fabrication practices for offshore platforms and, eventually, onshore structures all around the world. The loss demonstrated a pressing need to ensure the adoption of steel specifications and fabrication practices which would enable steel structures to operate in the northern North Sea conditions without the risk of collapse from the effects of a brittle fracture of a member. It is worth looking at the main points of the event. Originally built in the USA in 1952, *Sea Gem* was basically a steel pontoon about 100 m × 30 m in plan and 4 m in depth. It had been used in several different parts of the world until in 1964 the hull was substantially modified on two occasions at yards in France. At the time of its loss *Sea Gem* had been on station in the North Sea for some six months during which time it had been drilling gas wells. The catastrophe happened as the barge was being lowered by jacking it down on its legs prior to its being moved. Two tie bars made by gas cutting from steel plate fractured leading to a sequence of events which culminated in the barge capsizing; of the 32 men who were on board 13 died. A month earlier two similar tie bars had fractured without consequence and these had been replaced, apparently without any attempt being made to discover the reason for the fractures. Although the barge was working some 43 miles (74 km) offshore its operations were subject to the UK legislation applying on the Continental Shelf and the drilling was being performed under a production licence issued by the Ministry of Power. The UK Continental Shelf Act (1964) empowered the Minister of Power to make regulations which were to include provisions for the safety, health and welfare of persons employed on operations undertaken under any licence under the Act. A public enquiry into the loss was therefore set up by the UK Government. An investigation of the tie bars for the tribunal of enquiry[35] showed that they had poor fracture toughness and were of a steel susceptible to strain age embrittlement. The gas cut edges were irregular and in places had been repaired by welding and the fractures had originated 'from severe notches such as fatigue cracks, weld defects and the fillet radii between the spade end and the shank'. The consequent mode of failure of the remainder of the barge showed that the materials, the joint design and the fabrication practices were unsatisfactory for the service required of the barge.

The loss of the *Sea Gem* prompted the UK Government to introduce further legislation which came five years after the event in the form of the Mineral Workings (Offshore Installations) Act 1971. By 1973 only three of its ten proposed sections were in force and further sections included the provision for the Government to appoint an independent certifying authority to issue certificates of fitness for offshore installations. At a public meeting[36] convened by the Society for Underwater Technology in London in 1973 this proposal was strongly argued against by consulting engineers and others in statements such as: 'There are already insufficient data for the North Sea, and theories are as many in number as there are experts to expound them. Who, then, can be expected to say with certainty which one is correct? The certifying authorities? If the designer should disagree with the certifying body, who will decide between them?' It is interesting to recognise that at the time this meeting was held some large platforms had already been installed in the North Sea! In the event several certifying authorities were established and exercised their role success-fully.

9.2   The North Sea environment

Tradition held that round tubes were good for offshore platforms for two reasons: firstly, as structural members they were good in compression as they have a very effective and balanced distribution of material across the section which provides stability against elastic buckling. Secondly, round tubes present the same lateral resistance to the waves in the sea from all directions and this resistance is generally lower than would be given by a prismatic shape; associated with this their bending strength and stiffness is the same in all lateral directions. Significant hydrodynamic wave action is confined to a depth of water close to the surface of the sea, which includes the height known as the splash zone, the area of the structure intermittently wetted by the wave surfaces. Below and above the region affected by waves the benefits of the circular section are not as strong. Recognising this the Cleveland Bridge and Engineering Company Limited designed a large offshore platform in 1973 which they named the 'Colossus' (Fig. 9.1); its primary structure in the splash zone comprised circular tubes and the remainder was of flat plate construction. Castings were to be employed for

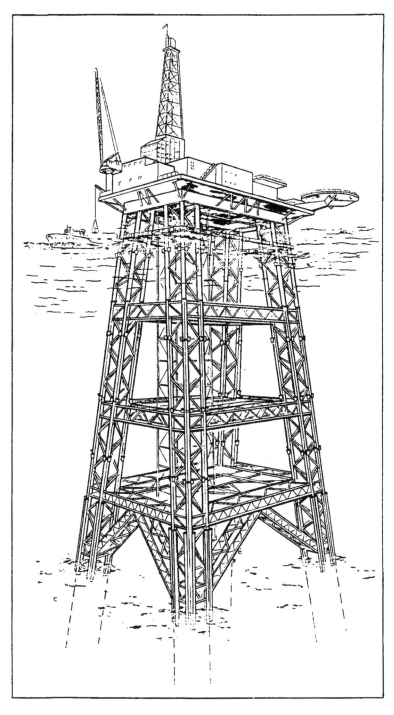

9.1 The Cleveland 'Colossus' deepwater platform (by courtesy of DCC Dixon).

complex joints thereby minimising stress concentrations so as to provide a long fatigue life. Arguments for that design, which was never fabricated, were that the lower structural members were made in the traditional rectangular box configuration commonly used in bridges so that secondary stresses were low and a long fatigue life was envisaged. In addition the structure employed well proven fabrication methods thereby reducing the uncertainties arising from novelty. Cast structural joints were not to become a practicality in offshore platforms until many years after and the 'Colossus' was ahead of its time in that respect.

It is interesting to see that in practice, where the legs of a tubular steel platform were particularly large, e.g. 10 m in diameter, the structural configuration often tended to be more conventional with external members passing into the leg with their loads being reacted by internal diaphragms. In some medium size tubulars it became the practice to introduce internal stiffening rings where the loads from bracing members were high.

What seems to have been in the minds of engineers in this matter of tubular joints was the experience that with small tubes there was no problem with welding the end of one tube onto the side of another. Familiarity with this detail perhaps persuaded them that this was a clean joint which one should not interrupt, although in some cases external gussets or rings were introduced to 'strengthen' the joint. These are not particularly desirable in the splash zone since they attract more hydrodynamic load than the clean tube profile and where they have been used it has been only on joints above the splash zone. There were proposals for joint configurations in which the brace was continuous through the chord thereby avoiding the generation of secondary bending stresses but such designs were not accepted by the offshore operators and their fabricators.

Was there at the back of the engineers' minds a resistance to the introduction of internal diaphragms from a fear of the sort of welding problems of the 1960s? The fabricators had only just started to overcome the problems of lamellar tearing in plate and chevron cracking in weld metal which were the bane of heavy fabricators in the late 1960s and early 1970s. However neither the set-on nor the set-through joint was more or less susceptible to these problems than others.

Regardless of all this what we see is a band of diameters and wall thicknesses of tube which attracted the set-on type of joint and which have the poor fatigue performance as evidence of the secondary stresses set up in this configuration. There was certainly a practical difficulty in welding internal structure into tubes of less than, say, 1 m in diameter. This was undertaken in some circumstances but with pre-heat temperatures of 100 ºC and higher. Welders had to wear a heat proof suit which made working very uncomfortable and welding very slow. But was this really the problem? Perhaps we shall never know. We might be forgiven for thinking that what the following story really tells us, with hindsight, is that some £10 million was spent finding out how to use poorly designed joints but perhaps that is being a little cynical.

Welded joints in the large offshore platform structure were considered to require special research for three reasons:

- There was little experience in the use of welded joints in thick carbon–manganese steel down to temperatures of –10°C where there were high local stresses such as those set up by the nature of the joints in tubular members then being designed. Fracture toughness requirements had been set down for process plant and for submarines, both of which used heavy plate, castings and forging. However the design details ensured that any local stress concentrations were modest. The material requirements for such products rested on a substantial body of work on fracture which had been instigated during the Second World War following fractures of a number of welded merchant ships. To cope with very low temperatures the process plant industry used alloy steels, austenitic steels or other metals such as aluminium.
- The fatigue performance of welded tubular joints in steel was not established in anywhere like the same degree of detail and confidence as for welded plate and sections. By the 1960s a number of countries were publishing standards containing fatigue classifications for welded joints, mainly for application to steel bridges. As described in Chapter 6, in the UK the British Standard Specification for steel bridges, BS 153, had carried welded joint fatigue classifications in its 1958 edition. Extensive research directed at the design of both civil and military bridges had led to the publication of more comprehensive data in the amendment to Parts 3B and 4 of BS 153 in 1962[15] and for welded aluminium alloys in BS CP 118 in 1969.
- The effect of a seawater environment on the fatigue life of welded joints in carbon–manganese steels was not well established, and nor was the effect of its corollary, corrosion prevention systems. The nuclear power industry had investigated the chemical effect of seawater on metals particularly in respect of their cooling systems, which all used seawater, but fatigue was not a mechanism which had been of concern.

9.3   The research

As a result of these concerns the UK Government decided to support an investigation of the engineering, material and fabrication needs of the proposed deepwater platforms. The Department of Trade and Industry though its Shipping and Marine Technology Requirements Board set up a committee called the Marine Materials Panel, which itself set up a subcommittee known as the Structural Steels Working Party. This working party was charged with the task of defining what relevant knowledge existed, what further knowledge was required and how this knowledge should be acquired through research programmes. During 1972 and 1973 this working party examined existing practices and research in progress in the use of heavy structural steel in other industries, particularly the process plant industry and from this developed the basis of preliminary recommendations for the design and fabrication of offshore steel structures. At the same time the working party commissioned The Welding Institute to undertake a study to identify areas of research which should be pursued to reduce the probability of fractures occurring in offshore steel structures in the North Sea. The conclusions and recommendations of that study were set out in a report[37] which was later published in summary form.[38]

The recommendations in that report lead to the drafting of a research programme, which came to be known as the UK Offshore Steels Research Programme (UKOSRP). This was to be conducted to examine the effect of seawater on the fatigue life of welded joints in steel and the fatigue performance of large welded tubular joints. The programme would also examine the fracture behaviour of the types of steel envisaged so as to be able to develop fracture toughness requirements of steels to avoid brittle fracture. The programme was to be funded by the UK Government and interests looked after by the UK Atomic Energy Authority as project manager. The fatigue and fracture testing was to be performed by The National Engineering Laboratory at East Kilbride and The Welding Institute near Cambridge with the support of the Harwell Corrosion Service of the UK Atomic Energy Authority. This programme was formally approved in 1973 and the design and manufacture of test rigs and specimens was put in hand. Progress was to be monitored by a new committee called the Offshore Steels Steering Group comprising representatives of classification societies, designers, operators, the British Steel Corporation and government departments. The initial advice of this group was that a significantly enlarged programme would be required to meet the agreed objectives. This was approved in 1975 by the Offshore Energy Technology Board which had by then subsumed the interests of the Shipping and Marine Technology Requirements Board in matters of offshore oil and gas exploration and production. The burgeoning development of the deepwater oil fields required that the programme be designed with the aim of providing immediate information on material and joint performance which could be applied to design in the short term as well as of producing basic knowledge and understanding for the long term development of the technology. The enlarged programme saw the introduction to the work of further contractors in the form of Lloyd's Register of Shipping Research Laboratory, Atkins Research and Development, the University of Nottingham and the

Springfields Laboratory of the United Kingdom Atomic Energy Authority.

Very quickly the programme expanded into a European programme with financial contribution from the European Coal and Steel Community and with co-operation between research laboratories in a number of countries particularly Norway and the Netherlands which had very strong domestic and commercial interests in North Sea oil production. International networks of researchers such as Sub-Commission XV-E of the International Institute of Welding and the Working Groups of CIDECT (Comité International pour le Développement et l'Etude de la Construction Tubulaire, Committee for the Study and Development of Tubular Structures) enabled more and more people to contribute further to the understanding of the particular characteristics of tubular joints.

The UK Government through its Department of Energy required that platform designs be examined by independent bodies. These were to be called Certifying Authorities who, if they were satisfied with the design and construction, could issue a certificate of fitness for the platform to be operated. These Certifying Authorities were private bodies initially constituted of consulting engineers with a range of disciplines. To provide a basis for the assessment of the integrity of designs of new steel platforms the Department of Energy commenced the preparation and publication of *Guidance on the Design and Construction of Offshore Installations.*[39] The results of the UKOSRP work were used to enhance this *Guidance*. In parallel with this were being developed national standards and codes of practice. In the United Kingdom the British Standard Draft for Development, DD55, was being prepared, eventually to become BS 6235 but which suffered the indignity of being withdrawn almost as soon as it was published for reasons which were not made public. The results of thousands of hours of intensive work by the members of the committee and their supporting staffs were nullified by this action. The Norwegian Petroleum Directorate published its own regulations, which for tubular joints closely followed the UK *Guidance*, sharing their research input and the results. In the United States API RP 2A section on tubular joints was expanded by including fatigue design data based on the results of the UKOSRP programme.

In the fracture toughness field the work, mostly conducted at The Welding Institute (now TWI), in the early and mid-1970s as an extension to their previous research programmes, would lead to CTOD criteria for steel and weld metals as used in offshore installations. Weld metals in particular required considerable development to offer adequate fracture toughness without resorting to exotic compositions which might in themselves create a corrosion problem owing to differential electrochemical potentials.

The requirements set down for the material and weld metal properties, particularly in the field of fracture toughness, were very demanding and could be achieved often only marginally by extremely close adherence to approved welding procedures which demanded very close limits on heat input. The requirements as interpreted by some North Sea operators and certifying authorities ended up placing a substantial cost and time burden on fabricators. CTOD testing was not then a regular service offered by test houses. It required sophisticated loading and measuring equipment and personnel trained and experienced in the testing techniques and the interpretation of the results. Exasperation in fabricators was induced by their having to produce PQRs (Welding Procedure Qualification Records) which for some joints ran to some 50 pages!

This was also the time when weld defect acceptance levels of a severity seen before only in power and process plant were being applied to complex three dimensional joints in thick steel plate when the only means of defect detection was ultrasonic examination, which at that time was not such a reliable technique as it became in the 1990s. Stories abound of lengths of welds in thick steel plate measured in tens of metres being excavated because of a reported weld defect which was found never to have existed. Small wonder that one well known and highly experienced Scottish welding engineer, exasperated after such an experience, was driven to say at a seminar in 1974: 'What I want is a weld metal so tough that it does nae matter if you can *see* through it!' In the event the improvements in weld metal and parent metal toughness over later years have almost made his wish come true for the consumable manufacturers in a relatively short time were able to offer weld metals which could well satisfy the requirements.

9.4   Platform design and construction

The contractual arrangements for conducting offshore projects were very different from the traditional civil engineering project. In this the client retains a consulting engineer, 'the Engineer', to design and supervise the construction. He is entirely responsible for its execution and success. In the offshore industry the major oil companies (or that subsidiary of an oil company assigned to 'operate' the field) frequently exercised very close control over the work of the designers who were in reality just design contractors. The design contractors then had no conventional responsibility for the specification and supervision of construction which was exercised by the oil company's engineering staff many of whom were extremely experienced. Much of the progress in materials and welding in the 1960s and 1970s was owing to their assiduous attention to the integrity of their structures destined for the northern North Sea. Whilst it may be invidious to define the contributions of individuals, it would be accepted by many that the energies of the late Harry Cotton, BP's Chief Welding Engineer at the time, moved forward many of the developments, driven by his determination never to see the like of the *Sea Gem* tragedy again.

9.5   Service experience

The record of structural performance of the larger fixed platform structures has been very good. A number of repairs have had to be made because of boat collision damage or damage due to dropped objects. There have been relatively few occurrences of early fatigue cracking in major fixed platform members; those which have occurred have tended to be in horizontal bracing near the water line which suffered wave action causing out of plane bending. Techniques of strengthening such joints *in situ* were developed and appear to have been successful. There appear to have been no brittle fractures in fixed platforms. Such a record is cause for some satisfaction bearing in mind the novelty of much of the construction of the early platforms and the lack of service experience. It might be said that the experience shows that designs were conservative, and if that is so then it must be seen as a good thing. The experience formed a sound base for developing lower cost methods of oil and gas extraction. How much better than a scenario of having early disasters and then having to improve designs.

10.1 Basic requirements

It is generally accepted that any business activity must be conducted within an overall discipline which ensures that what is needed or wanted is defined and that actions taken to satisfy that need or want are put in hand and executed in an effective manner and at the appropriate time. This requires three basic resources:

- people with the necessary knowledge and skill
- facilities to enable those people to exercise that knowledge and skill
- inputs or materials which can meet the requirements of the job.

With these resources it is then necessary to have:

- a plan of what activities are to be pursued with which material and when
- means for conveying instructions to the parties and individuals involved
- means of controlling activities and/or of demonstrating that their outputs conform to the requirements.

These features will be recognised as having been part of engineering for as long as it has been pursued and so there is nothing new here.

10.2 Contracts and specifications

A commercial contract is usually an agreement between two parties whereby one party supplies to or does something for the other party for a *consideration*. Such a consideration is usually a payment in money or some other negotiable device. What has to be supplied or done, in other words, what is to be bought, purchased or procured, whichever word is used, needs to be spelt out clearly in a description. For simple purchases this description may simply be the name of a proprietary article and common business practice is to enter into a simple agreement on the basis of a *purchase order*. This will simply ask that a certain quantity of an item in the catalogue be supplied for a price within so many days or weeks. There will be standard conditions of the supply which may be those of the buyer or the supplier or a negotiated compromise. Usually the description of the product is that of the supplier as it appears in his catalogue or brochure. Such a method may be used for more complicated engineering products but only where that product is made of standard items. An example is a simple reaction or separating vessel with various nozzles, all made to standard specifications. The purchaser describes what is required in a drawing or sketch calling up standard parts. For more complicated and one-off items such as a building, crane or power station, the purchase order is an inadequate instrument. There are many stages in the design and construction of such a project and the money is paid out in accordance with the work done. Evidence of achievement and verification of the quality of the work has to be allowed for as do procedures for addressing disputes. The product is a one-off and needs to be described in a specification; there is no prior product to use as a sample or model. At this point we may see a divergence in the approaches to specification writing. At one extreme the specification says only what the item is required to do; this is called a *performance specification*. For example, a stockyard crane to lift x tonnes to a height of y metres and carry it for a distance of z metres; this will be accompanied by the details of the site. The customer leaves it entirely to the supplier to design, manufacture and install the crane and make it work. The stockyard owner may know nothing of the design of cranes or even the laws and regulations surrounding their construction, although he will have to learn about those relating to the use of cranes. He can ask only for what he needs. The other extreme is where the client describes the item in great detail, even to the extent of providing conceptual or detail designs from which the supplier may need to develop working drawings. As an example, a specialist chemical manufacturing company may design its own processes and its chemical engineers know exactly what their plant needs to be. They will write a specification in great detail. Which approach is used depends greatly on the nature of the client. In practice such extremes are rarely followed for a number of reasons. The crane will require foundations for the rails which will require knowledge of the soil conditions and it will require an electrical power supply. The client will then have to retain specialists to deal with those matters. They will have to agree requirements with the crane supplier. In the end the client may find it easier to employ a consulting mechanical/electrical/civil engineer familiar with cranes to act on his behalf. The engineer's experience will tell him that there are certain things about cranes which have to be specified in detail for that yard because the standard items, although acceptable, are not the best. At the other extreme the chemical company may get what it asks for but the price may be high because the specified work is not the way that the suppliers usually go about their work. In addition there is a danger for the client that his detailed specification is deemed to be an instruction to the supplier which, if it turns out to cause a problem, may in law place the responsibility for any consequent losses or damage on the client.

The trends over the past few years to partnerships rather than traditional adversarial agreements helps to avoid these two extremes but it is still essential that an agreed specification for the works be derived as early as possible. The use of an effective management system will ensure that the intent is realised.

10.3   Formal management systems

Most clients and customers have become less and less tolerant of late deliveries and of products which do not perform as required or as expected. A major move to reduce late and poor performing products by formalised management approaches was made in the 1950s particularly in the context of procurement for the defence and energy industries. The causes and sources of delays and poor product performance were analysed and actions set out as a management system which would prevent these happening. These formed the basis of various documents. The Canadian Standard, Z299, initially issued in 1975, described a system prepared for nuclear power station construction. This was particularly suited to site construction as opposed to factory manufacture. In the UK, BS 5179, based on defence standards, was issued as a guide to the evaluation of quality assurance systems but was withdrawn in favour of BS 5750, Quality Systems, which was first published in 1979, and apparently directed mainly at manufacturing. This set out the most important features, or elements, of a management system which, when operated within a contractual situation between a purchaser and a supplier, would lead to design and production of engineering products conforming to a specification. After a period this standard was used to form the basis of the ISO 9000 series of documents, which have seen various amendments and additions in attempts to apply the original system elements to non-engineering and unspecified products and services and to other contractual, or even non-contractual, situations for which they were never intended. However within the scope of this book we do not need to address those points.

It is generally agreed that the achievement of and demonstration of consistency of conformance of the product to the specification is an essential requirement of any industrial process. In many manufacturing processes part or all of the operations may be performed by mechanical devices. These devices can be designed to include their own control systems so that with adequate maintenance and with the correct input of materials, barring external disturbances, they can be used to manufacture products to a high level of accuracy and precision. There are claims by some that such practices render final inspection unnecessary whilst others would not be so bold. At the raw material and semi-finished materials stages the conformance of the product can be assessed readily by tests on samples taken from production. Once the manufacturing operations have moved along to making the final product, such destructive testing is no longer always a viable approach to control except for low unit cost or mass production items which can still be sampled. Even so, post manufacturing inspection has long been considered an inefficient approach to quality control for it detects non conforming items after anything can be done to correct them. The result is that the parts themselves are wasted as will be others made at about the same time. A more effective approach is to control the process so that the parameters which affect the conformance of the product are maintained within the limits which have been shown to produce conformance.

10.4 Welded fabrication

Formal management systems[41] have been applied to welded fabrication activities for longer than most. A welding procedure, or to be more precise, a welding procedure specification, is more than what is generally understood as a procedure. It is a statement of the whole input to the manufacture of a welded joint. It defines the material(s) to be joined, the welding process, any welding consumables, edge preparations and welding position. Also included are welding conditions which in arc welding mean quantities such as welding current or wire feed speed, voltage and electrode run out length or welding travel speed. Preheat temperature, sequence of weld runs, any post weld heating or heat treatment, and non-destructive testing are other major points in a welding procedure.

Much welding is still done manually which gives a potential variability in the results as is the case with any process operated by the hand of man. In a manual welding operation the welder, the materials and the equipment are all part of the manufacturing system. To acquire confidence that the product will conform to the specification the welder, the equipment and the procedure to be used must be confirmed as being capable of making the required product. This confidence is acquired by checking that the equipment will be working within its operating limits and by giving the welder a trial joint to make which requires the same skills as the actual job. This trial joint may then be non-destructively examined by radiography or ultrasonics and then cut into pieces, some for metallographic examination and hardness tests, others for mechanical tests of strength and ductility. Sample joints may have been made with the welding procedure to confirm that it is capable of producing the required joint characteristics.

To provide prior evidence that the welders and the welding procedures are capable of providing the required fabrication and so avoid the need for tests for every job, records of welder approval tests and welding procedure approval tests are kept by the fabricator. These records are maintained as certificates signed by the testing and witnessing authorities. Depending on the interests of the parties involved in the job, welder tests and welding procedure tests may be witnessed by the customer or some independent third party on his behalf. There are some industry wide schemes in which independent surveillance is undertaken. This is intended to provide confidence that a manufacturer's personnel, equipment, organisation and operation are such that customers can accept that the manufacturer is competent to do the work. This replaces separate tests for each customer which they would have to pay for themselves as part of the contract. Such schemes have their limitations and the display of a certificate of conformance to some management system standard does not represent any guarantee that a firm or an individual will perform as required in any particular situation.

As in any other business the production of welded fabrications requires educated, knowledgeable, trained and committed people working within an appropriate management system. In summary, basic confidence in the welded fabrication of a viable design is achieved by having:

- competent welders as demonstrated by welder approval certificates
- relevant welding procedures as demonstrated by welding procedure specifications which have been tested
- competent inspection personnel as demonstrated by relevant certificates.

Mechanised or robotic welding operations still require the input of a knowledgeable, skilled and qualified operator and welding procedure specifications still have to be prepared and verified.

11.1 Weld defects

Why have defects? A cynic might observe that there can be no other branch of manufacturing technology where so much emphasis is laid upon getting things wrong and then attempting to justify it as there is in welding. 'Weld defect' is a term almost inseparable from welding in the minds of many welding engineers. It seems to be an essential part of the welding culture that weld defects should be produced! What is commonly meant by a weld defect is some lack of homogeneity or a physical discontinuity regardless of whether it diminishes the strength or damages any other characteristic of the weld. Welding is one of the few final manufacturing processes in which the material being worked on exists simultaneously at various places in two phases, liquid and solid. This together with the large temperature range and the high rates of change of temperature gives rise to the potential for great variability in the metallurgical structure of the joint and its physical homogeneity. It is as well to recollect that discontinuities in materials are not necessarily undesirable. Indeed the strength of metals depends upon their containing dislocations on an atomic level, i.e. disturbances or what might be called in other circumstances 'defects' in the regular lattice of the atoms, without which metals would have very low strength. It has to be recognised also that weld discontinuities, be they lack of fusion or penetration, porosity or cracks, do not necessarily result in a defective product in the legal sense that it is not fit for its stated purpose. Attempts have been made to get round the situation by calling these features flaws, discontinuities or imperfections but the word *defect* remains common parlance even though it is really rather irrational. We would not call for a polished finish on a steel bar if an as-rolled or rough turned finish were adequate for the job it had to do; we do not call the latter surfaces defective, we call them fit for their purpose. This concept of fitness for purpose began to gain acceptance in respect of welded products in the late 1960s when Harrison, Burdekin and Young[42] showed that the commonly used weld defect acceptance levels in some standard specifications and codes of practice were extremely conservative and had frequently led to unnecessary repairs and rework during fabrication. Such repair work was not only an added cost but also delayed fabrication and even whole construction programmes. Furthermore the conditions under which repairs had to be conducted resulted in the repaired weld sometimes being of poorer performance than the original so-called defective weld. The methods of fracture mechanics were developed to make it possible to review the effect of a weld defect in terms of the service requirements within the process known as engineering critical assessment (ECA).

Within the whole discipline of metallurgy, welding metallurgy is a specialised subject which is distinguished by having to address the behaviour of metals where there were:

- high rates of temperature change
- high temperature gradients
- changing solubility of gases in the metal
- small volumes of metal rapidly changing from solid to liquid state and back again
- transfer and mixing of metals and non-metals in a complex gaseous and electrical environment.

These features set up rapidly changing fields of strain and the resulting stresses add to the physical, metallurgical and mechanical characteristics of welds.[43]

The features which are called weld defects can be attributed broadly to two main sources – workmanship and metallurgy. Workmanship defects are those where the skills of the welder have not matched the demands of the weld configuration. Examples of such defects are lack of penetration, over-penetration, lack of fusion, undercut and poor profile. Metallurgical defects[43, 44] arise from the complex changes in microstructure which take place with time and temperature when a weld is being made. They can arise from the natural composition of the steel or from the introduction of extraneous substances to the metal matrix which is incapable of absorbing them without the induction of high strain and consequent fracture. Examples of such types of feature are hydrogen induced cracking, hot cracking and lamellar tearing. Notwithstanding their different origins the occurrence of each of the two groups of defects can be avoided by proper management of the fabrication operations (see Chapter 10).

Bearing in mind the subject of this book we need to see if the origins of weld defects can lie in the nature of the design. If we accept that it is the designer's responsibility to specify both the materials and the configuration of the product then we start with the materials. An essential part of the specification is that the material shall be weldable whatever we take that to mean. It is at this point that this concept of the 'designer' starts to run into trouble for we know that the selection of the material is intimately bound up with the choice of the welding process and the welding procedure and vice versa. Similarly the weld preparations will be decided not only by the type of weld determined by operating requirements of the joint but also by the welding process, the position in which welding is done and the sequence of assembly of the fabrication.

It is for this reason that design drawings may be followed by fabrication drawings followed possibly by shop drawings which will themselves call up the welding procedures or summaries of them called data sheets. Design in these circumstances is an iterative process converging as quickly as possible to a solution which meets the project requirements. From now on if we speak of a designer we are really referring to one of a number of parties with different roles. As we saw, weld defects occur from two main sources between which there is some interaction.

11.1.1 Some common workmanship based defects shown in Fig. 11.1

11.1.1.1 Lack of sidewall fusion

The arc fails to melt the parent metal before the weld metal touches it. The molten weld metal rests against the parent metal without fusing into it.

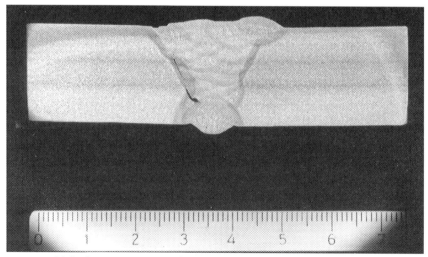

11.1   Some common workmanship based defects (photographs by courtesy of TWI). (a) Lack of sidewall fusion.

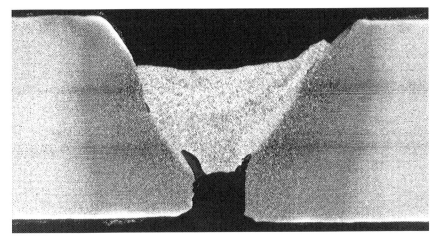

11.1 (b) Lack of penetration.

11.1 (c) Undercut.

11.1 (d) Poor profile.

This can occur because:

(a) the arc is not hot enough for the thickness of the metal
(b) the arc travels too quickly along the joint
(c) the arc is not directed at the parent metal
(d) the weld metal flows ahead of the arc, preventing it from impinging on the parent metal.

Only (c) is within the scope of the designer's influence if an inaccessible joint or an unsuitable weld preparation prevents the welder from directing the arc at the edge preparation.

11.1.1.2 Lack of root fusion

The arc fails to melt the metal at the root, causing a similar condition to that in lack of sidewall fusion.

The causes are similar to those of lack of sidewall fusion.

11.1.1.3 Lack of root penetration

The weld does not reach through the full depth of the preparation.
This can arise because:

(a) the root gap is too small for the welding conditions
(b) the root face is too large for the welding conditions
(c) the welder may not be sufficiently well practised or trained in the technique, particularly in positional welding, as in pipes.

11.1.1.4 Undercut

The parent metal is washed away adjacent to the weld.
This can arise because:

(a) the welding current is too great for the welding position
(b) the welder's technique encourages washing out of the parent metal.

11.1.1.5 Poor profile

The weld surface is erratically shaped, peaky, underfilled or overlaps the parent metal.

Mainly caused by wrong welding conditions, lack of welder skill, practice or diligence.

11.1.2 Some common 'metallurgical' defects shown in Fig. 11.2

11.1.2.1 Hydrogen induced (cold) cracking

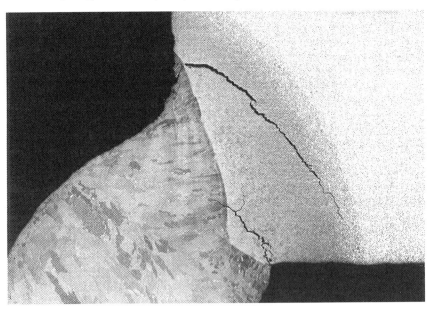

11.2 (a) Some common metallurgical defects (photographs by courtesy of TWI).

The occurrence of cold cracking in steel is a function of both microstructure and hydrogen dissolved in the metal; in simple terms it can occur if the microstructure is very hard, usually in a martensitic heat affected zone, and the dissolved hydrogen level is too high for this hardness. It is called cold cracking because when it occurs is when the metal has cooled to ambient temperature. This type of cracking can also occur in the weld metal but this is less common. It is prevented by two measures which are part of normal good practice in the design of welding procedures:

(a) ensuring that a combination of pre-heat and welding heat input is used so that the rate of cooling of the heat affected zone is not so high as to quench it to a high hardness

(b) minimising the amount of hydrogen taken up by the steel by ensuring clean metal surfaces (no grease, paint or moisture) and using low hydrogen welding consumables.

In higher alloy steels it is sometimes impossible to reduce the hardness sufficiently and post (weld) heating is applied. This allows hydrogen, which might otherwise cause cracking, to diffuse out of the steel over a period of hours. In marginally hardenable steels the same effect is achieved just through preventing the steel cooling down quickly after welding by covering the work with heat-proof blankets.

11.1.2.2 Hot cracking

Hot cracking can occur in the heat affected zone as liquation cracking when on being heated by the welding arc non-metallic substances in the steel (usually sulphides) melt whilst the steel is solid and form layers of weakness which fracture under the thermal stresses of welding. In weld metal this form

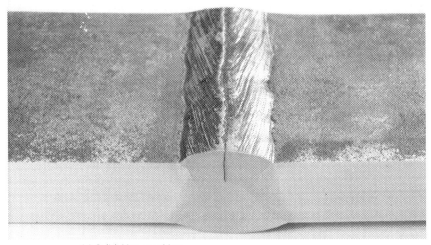

11.2 (b) Hot cracking.

of cracking is known as solidification cracking; as the weld metal cools down the steel solidifies leaving the non-metallics still liquid. This has the same result as liquation cracking. The pattern of its appearance can be influenced by the freezing pattern of the weld metal, sometimes appearing along the centre line of the weld as the last area to solidify after the metal crystals have formed in a single run weld. These forms of cracking are prevented by attention to the sulphur content of the steel and weld metal. There is a broad requirement in single run butt welds that to avoid the circumstances where weld metal hot cracking might be a problem it is customary to restrict the depth to width ratio of the weld.

11.1.2.3 Lamellar tearing

Microscopic islands of sulphides and other compounds are produced in some steels and when the steel is rolled into plates these islands become platelets or lamellae Those near the surface can be melted by the heat of any welding and if the combination of heat and thermal stresses is sufficient

11.2 (c) Lamellar tearing

these lamellae will become points of weakness and allow the steel to fracture in a rather woody looking way known as lamellar tearing. The incidence of lamellar tearing can be increased by hydrogen cracking which can act as an easy starting point for a tear. Lamellar tearing is most likely to occur under T joints in steel which has been rolled down sufficiently thin to produce lamellæ but which is still thick enough, in combination with the other parts joined, to restrain the incipient thermal contractions which can set up high stresses at right angles to the steel surface under the weld. In practice this tends to mean steel plate in thicknesses between 20 and 50 mm. The risk of lamellar tearing can be avoided by using plate which has been processed in such a way that it does not contain the lamellae. This is achieved either by modifying the shape of the inclusions so that they do not cause planes of weakness or by removing the material of which they consist. The latter approach was found to encourage the occurrence of hydrogen cracking; previously the non-metallic inclusions had acted as 'sinks' for hydrogen drawn into the steel and without them the hydrogen entered the steel matrix and caused trouble. The level of resistance of a steel to lamellar tearing is conventionally indicated by the observed reduction in cross sectional area of a tensile test specimen taken in a direction at right angles to the surface, the Z direction as it is known. Steelmakers offer steels giving various guaranteed levels of reduction in area, e.g. 15%, 20% or 25%. In practice, a steel from a reputable maker can give values of up to 75%. Which level is chosen for a particular circumstance may come from specialist experience or an application standard.

11.2 Quality control

11.2.1 Quality in welded joints

The means of the control of quality are many and varied but they are all directed at ensuring that the product meets the specification. Specifications can be very tightly defined or they can be very loose; they may deal with many characteristics of the product or just a few. A phrase commonly found in more traditional specifications is *good workmanship*; this has no measurable meaning and so is very subjective. It is taken to mean something made in a way which has become commonly accepted in the industry as achievable by a trained and experienced workman and generally meets its purpose. Many such workmanship criteria are very old and incorporate sound techniques developed as a result of perhaps centuries of experience. The use of some techniques however is based on a misunderstanding of everyday observations; an example is the oft heard explanation that one heats a steel plate before welding with a gas torch 'to drive out the moisture'. This is arrant nonsense and arises because people seeing the water vapour of combustion in the flame condense on the cold plate take it that the water must have come out of the plate. Confirmation that the product meets a workmanship requirement lays with the opinion of the person charged with examining the product for acceptance, the inspector. This approach can work quite well in an industry with a stable workforce making similar products repeatedly over a long period. Judgement is made on the basis of past satisfactory performance and acceptance by the customer. In the current world of more fluid workforces and where there is less tolerance allowable on the product performance, it may be necessary to define the acceptability of the product by certain measurable parameters. This may be assisted by the comparison with samples or replicas of an acceptable product. Both of these methods of judging the acceptance of the product have the drawback that they examine the product after time and money has been spent on it. If the item is unacceptable it has to be rejected altogether or it may be repaired both of which actions represent a waste of resources and money. Statistical analysis of inspection results in mass production can detect trends away from the desired product characteristics and the equipment can be adjusted to correct this trend. Much welding work does not lend itself to that approach which relies on identifying discrete items of production. In particular, manual welding of a long seam has to be completed before it can be inspected. Any 'defects' then have to be excavated and re-welded. Such work if it is to be free from interruptions needs the attention of trained and competent welders and well designed joints which do not require unusual feats of skill to weld.

When welding with mechanised equipment the welding conditions can be set up and automatically, or even robotically controlled during welding. The truly adaptive system will make allowances in welding conditions for variations in fit up as they may affect the root penetration or alternatively the penetration will be monitored and the welding conditions adjusted accordingly. This is the basis of process control, used in many industries, as a means of ensuring that the output conforms to the specification. In a perfect world, post weld inspection would then be unnecessary; few of us would have such confidence or relinquish the opportunity of passing an eye over the completed work. It is easy to concentrate hard on the measurement of detail and forget to check that all the welds are in the right place or even there at all!

In welded fabrication weld defects are not the only subject for quality control. Dimensions and materials are also important to the quality of the final job as in all engineering work but both are of particular concern in welded construction.

The concept of a dimensional tolerance is well established and such tolerances represent a band of dimensions based on a specified nominal figure. These dimensional tolerances are necessary for a number of reasons. A prime reason is that it is impossible to make anything to an exact dimension and having made it to be able to measure it exactly. A batch of nominally similar items cannot all be made exactly the same and the magnitude of the tolerance which has to be allowed to account for differences between one nominally identical item and another is a measure of the precision of the manufacturing process. The smaller are the tolerances allowed the finer has to be the capability and control of the manufacturing processes which implies cost, a reason for making tolerances as wide as is possible. Clearly the tolerances cannot be so wide as to prevent mating parts from fitting properly, e.g line-up of bolt or rivet holes, fit-up of parts to be welded. Buckling strength may place a limit on the flatness of plates and straightness of columns. Tolerances are applied to shafts and other round parts which have to fit into holes. One end of the scale of such tolerances may have to give a running fit to allow rotation or sliding and the other end an interference or force fit to connect parts firmly.

As an example, the dimensional tolerances on the dimensions of the parts of a machine tool may be set for reasons such as:

- to give the machine its correct overall dimensions
- so that individual parts will fit together and will be interchangeable
- to provide a seal between parts containing fluids
- to give the required degree of fit, which may be a force fit for fixed items or a running fit for journals/bearings
- to transfer loads uniformly
- to provide balance in rotation.

In steel fabrications tolerances are placed on dimensions to recognise that it may be impossible to fabricate beams and columns which are exactly straight or plates which are exactly flat or cylinders and domes which are exactly to the required shape. The dimensional tolerances are set to avoid instability or to keep secondary stresses to within defined limits. In aircraft, ship and car body manufacture tolerances are placed on material thickness to control weight and on aircraft and ships tolerances on dimensions to provide in addition to the above features the necessary aerodynamic and hydrodynamic performance.

In welded fabrication sources of dimensional variation are thermal distortion and residual stress. The welding arc is a point heat source at a very high temperature which moves along the joint. The sequence of heating and cooling which takes place leads to expansion and contraction of metal through a range of temperatures and strengths. The result is that in some circumstances there remain locked in stresses, called residual stresses, and in others distortion from the original or desired shape of the part (Fig 11.3).

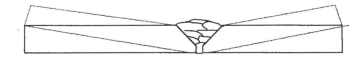

11.3 Distortion in a welded joint.

Even before a welding operation, parts may have to be set so as to neutralise the distortion which is expected. A simple weld bead on a thin plate will demonstrate how thermal distortion manifests itself. A multi-run weld made from both sides introduces a more complicated sequence of heating and cooling which will have its own effects. Welding is not the only cause of distortion in members. Rolled sections such as universal beams contain residual stresses owing to the different thicknesses of the section cooling at different rates after it is rolled. This is of little consequence when the section is used complete. However it is sometimes convenient to slit a section along its mid-line to make two T sections. Very often such a slitting operation will release the balanced residual stresses and the section then adopts a curved shape. Distortion can be experienced in other types of construction than welding. Riveted aluminium alloy structures as used in airframes can distort under the build up of the local strains introduced by each rivet setting. The sequence of riveting has to be planned to minimise this type of distortion. In some fabrications distortion may be accepted, as can be seen in many ship hulls. Distortion can appear during machining as stressed layers are removed and even in service as residual stresses are redistributed with time or by the effect of service loading. This is undesirable in some products and particularly in machine tools or fixtures where dimensional accuracy and stability are of the utmost importance. To avoid such in-service distortion it is usual to thermally stress relieve steel fabrications before machining or even at intermediate stages. This stress relief is achieved by heating carbon–manganese steels to some 580–620°C, and holding at that temperature for a time depending on the thickness of the steel. This relaxes the stresses to a degree but will not eliminate them altogether. A less frequently used method is to vibrate the fabrication through a range of frequencies which will locally yield out high residual stresses. Completion of the process is signalled by the reduction to a constant level of the required input energy from the vibrator, akin to a cessation of hysteresis. This treatment does not affect the microstructure of the steel and so does not offer the same benefits as heat treatment where improved resistance to brittle fracture is required.

It is basic to the engineer's role to recognise that it is either impracticable, unnecessary or not cost effective to define not only dimensions but mechanical properties and other characteristics of materials or structures to exact levels of accuracy or precision. Tolerances may be based on the ability to perform a measurement, on the consistency of raw material supply, on the capabilities of manufacturing processes or on the performance requirements of the structure in relation to the cost of manufacture. Steel may be made from raw materials or scrap, both from a number of sources, and the skill of the steelmaker is to end up with steel of a composition which meets a specification. Clearly handling and mixing tonnes of white hot liquid metal makes exact control of composition difficult and so tolerances are placed in steel specifications not only in terms of their chemical composition but their mechanical properties.

11.2.2 Inspection methods

11.2.2.1 Visual inspection

It might seem so obvious as to not require description but this is a key method of inspection without which the other methods are blind. It requires a qualified and experienced welding inspector not only to observe a completed weld but to be able to diagnose the conditions which have led to its condition. The visual inspection will reveal such features as surface breaking porosity, undercut, cold lapping, lack of fusion, cracking, lack of penetration, over-penetration, poor profile and, perhaps even more important, the complete absence of a weld, which is not unknown.

11.2.2.2 Magnetic particle and dye penetrant

There are two ways of revealing the presence of certain features at the surface of a metal which otherwise would be too fine for the naked eye to detect. In the magnetic particle method the metal (it must be a ferritic steel) is locally magnetised; discontinuities such as cracks at or near the surface concentrate the magnetic field which is then shown up by iron powder or a suspension of iron powder in a liquid sprinkled or sprayed onto the metal and which is attracted to the concentrated field. The dye penetrant method is used on non magnetic metals such as aluminium and stainless steel. A strong dye is sprayed onto the metal and soaks into any cracks or other surface breaking gaps. The dye remaining on the surface is wiped away and the surface is then sprayed with a fine chalk emulsion. This will draw up any dye near the surface so that the position of cracks and so on will show up as coloured lines or patches in the white chalk.

11.2.2.3 Radiography (X-rays)

Radiation passing through an object strikes a sensitive film giving an image whose density depends upon the amount of radiation reaching the film. This will show up the presence of porosity or cracks in a weld as well as variations in weld surface profile. Hollows in the weld surface will show up darker than the rest; cracks and pores will show up even darker. A crack which lies in a plane parallel to or close to that of the film will not show up well (Fig 11.4). The source of the radiation may a be an X-ray machine or, for site use, a radioactive isotope. The method requires that both sides of the subject be accessible, for the film on one side and the source on the other. The film is developed like a photographic film and has to be viewed in a specialised light box. The film can be stored for as long as is necessary.

11.2.2.4 Ultrasonics

A beam of high frequency sound is projected from the surface into the metal; the echo from the opposite face or any intervening gap is received back and the time between transmission and reception is measured electrically and displayed on an oscilloscope screen (Fig 11.5). A trained operator can tell what size of feature a signal or 'indication' represents and where it is. There is no record of the examination except that recorded in writing by the operator.

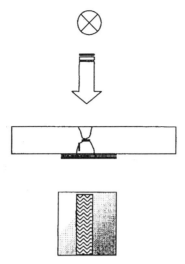

11.4 Radiograph of a butt weld. The lack of root fusion is not as clearly revealed as the lack of sidewall fusion.

Signal from defect

11.5 Ultrasonic examination of the butt weld.

11.2.3 Extent of inspection

Traditional mass production relied for quality control by either inspecting every item during and/or after its manufacture or, to save time and cost, inspecting a sample. In some products the sample may have been tested to destruction or cut up to confirm its conformance to the specification. Statistical techniques are used to define the rate of sampling or the sample size to achieve the required level of confidence. If the samples show that over a period of time a product characteristic, e.g. diameter, is moving towards a tolerance limit, perhaps because of tool wear, the machine can be re-set to give a diameter at the other end of the tolerance band. In more sophisticated circumstances the actual parameters of the process will be continuously monitored and adjusted to keep within the limits which will give products within their own limits. Clearly such sophisticated methods can be applied to discrete mechanised welding operations, such as resistance welding. Their application does of course require that the parts to be joined are themselves controlled in thickness and fit-up. Sampling is valid only where there is a basically repetitive manufacturing operation; in statistical parlance each item must come from the same population.

Owing to variations in material composition, fit-up, arc length and so on, manual or mechanised arc welding may produce defects which, except under conditions of gross malpractice, appear scattered on a more or less random basis. This is very different from the gradual deviation of a dimension from a nominal size. The practice of sampling is often to be found in specifications for arc welded fabrication unsupported by any statistical basis. For example the specification may say that '10% of welds will be radiographed'. This may give results of some significance if the welding is mechanised so that it can be said to be all of one population; even so there needs to be clarification as to whether this is intended to mean 10% of welds, i.e. one weld in ten, or 10% of welding, i.e. one tenth of each weld length or one tenth of the weld length chosen in a random manner. Many such specifications fail to say what action is to be taken if unacceptable features, 'weld defects', are found in the 10%. Certainly most of them say that such defects must be repaired. This then leaves the question as to what happens to the remaining 90% of the weld for if there is a defect in 10% there is some probability of there being defects in the rest. Some specifications deal with that likelihood by requiring that all remaining welds be 100% examined until the cause of the defects has been ascertained and resolved. Others seem to leave that possibility unconsidered. Of course even if no defect has been found in the 10%, there is perhaps still a chance that there will be defects in the remainder. Some specifications try to keep manual welds divided into populations by calling for '$x\%$ of each welder's welds to be examined'. However the discovery of a defect in one weld in a manual process may not be related to the remainder of the same welder's welds and, again, no action is usually defined. The best that can be said about such approaches is that they detect gross malpractice and that they may create an environment in which welders know that there is a chance of any defects being discovered. The worst is that they do not offer any true level of confidence that the work conforms to the specification. In neither case can this be said to be a satisfactory way of going about quality control yet these clauses are still to be found in specifications for important structural works. The drive to avoid 100% examination derives from a desire to save costs. However it has to be recognised that this is not always the cost of inspection, which may be quite trivial, but in the desire to avoid greater costs through having to repair the defects which might be found by 100% examination.

Project specifications often reveal a woeful ignorance of welding and non destructive testing in their compilers. Typically: '... butt welds shall be ultrasonically examined. ... fillet welds shall be examined by magnetic particle (MPI) ...' The inference is that these are alternative means of examination. This is patently not so, since both types of weld require visual examination and MPI to detect surface defects. Ultrasonics can be used to detect sub-surface defects but this is feasible only on butt welds. Fillet welds are an unsuitable subject for ultrasonics except by specialised means and should be excluded from ultrasonic examination purely on feasibility grounds and not because MPI is a substitute.

Inspection levels must be related to the nature of the product, the method of welding and the overall structure of the management system. Quality control is most effective when exercised on the inputs rather than on the outputs.

11.3 Welded repairs

The consequence of finding a fatigue crack or other type of damage will be a need to decide whether to scrap the item, repair it or use it as it is. This decision will depend on the cost and feasibility of repair against the supply of a new item. A question may be raised as to whether a repaired item will last as long as the original. To some extent this will depend on the accessibility of the damaged area and so the quality of the repair which can be made. If a fatigue cracked full penetration weld is replaced with a partial penetration weld because of lack of access then the repair is unlikely to give the longevity of the original; it might indeed just be a waste of time and effort to attempt the repair.

As in all welding the welders must be suitably qualified and the whole repair procedure must be planned in detail and, where necessary, verified by testing. For some large and costly plant it has proved justifiable to create a replica or mock-up of the joint area on which to develop special purpose equipment and techniques and to carry out trials before embarking on the actual repair. In repairing a weld there is no reason to believe that the repaired weld need be inferior to the original. Welded repairs in originally unwelded areas will, of course, have the characteristic of a welded joint in that material. The repair must commence with removing all the damage, confirmed if necessary by appropriate non destructive examination methods, and any associated distortion corrected. On some types of material such as alloy steels it may be desirable to remove all of any existing weld and heat affected zones as their properties may be affected by multiple thermal cycles. As for all welding work, the surfaces to be welded and adjacent areas must be cleaned of any paint or other substance which has accumulated during service or during any recovery operation. Suitable edge preparations must be designed and executed and any arrangements for pre-heat installed. Welding can then proceed as in the procedure and any interpass temperature and post weld cooling rates controlled. Post weld inspection will then be conducted, preferably before any post weld heat treatment. This inspection will require special planning if an elevated post weld temperature has to be maintained until post weld heat treatment is started. For *in situ* repairs the size of the item or restricted access may require a local heat treatment. Equipment giving control over temperature gradient/time as well as temperature/time may then be required. Final inspection will take place after an agreed time after the item cools to ambient temperature. This is a common procedure with steels to allow time for any delayed cracking to occur. This practice arose from past experience, with thick steels in particular; instances had occurred when a fabrication had been inspected and passed but was later found to contain cracks. Hydrogen induced heat affected zone cracking in steels, also called cold cracking, is known to occur some hours, or even days in thick sections, after cooling on cessation of welding. Some parties held that this could have been because the inspectors had missed the cracks in the first place, which may have been true or just uncharitable!

11.4 Engineering critical assessment

Looking back at man's recent use of iron and steel we find that in 1854 William Armstrong embarked on the design of a rifled gun as a replacement for the cumbersome field guns which at that time fired cast iron balls. James Rendel, famous for his work in civil engineering, encouraged Armstrong to consider steel in place of wrought iron, a transition which had commenced in civil engineering some years previously. Armstrong agreed that steel, having a much greater strength than wrought iron, should be the better metal for standing up to the pressure in the barrels. However experiments convinced him that in resisting explosive loads tensile strength was not the correct criterion. He therefore adopted the technique used in manufacturing sporting guns whose barrels were made 'by twisting long slips of iron into spiral tubes and then welding together the edges by which means the longitudinal length of the slips becomes opposed to the explosive force'. Armstrong had a great rival in the person of Joseph Whitworth who proposed a gun made from large forgings, in a total contrast to Armstrong's slip method. Armstrong's comment was: 'To make large guns on the principle of solid forged tubes either of steel or iron I consider entirely out of the question, because we can never penetrate the interior of the mass so as to discover the existence of flaws.' Alfred Krupp in Germany, of the second generation of that dynasty, was of course a natural rival of Armstrong and

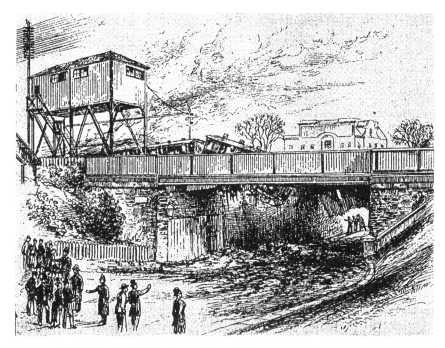

11.6 Structural failure in a railway bridge.

two years younger than him. In 1863 Armstrong wrote to Stuart Rendel, one of James Rendel's three sons and Armstrong's manager in London, reporting that one of Krupp's guns had burst, '... flying into a thousand pieces. All the fragments were sound so that the failure was purely due to the intrinsic unfitness of the material.' (Stuart's brother George was manager of the ordnance works of Sir W G Armstrong & Company and his other brother, Hamilton, was responsible for the engineering design of Tower Bridge in London.)

As we saw in Chapter 2, cast iron was known to be susceptible to fracture and there were a number of instances of catastrophic failure of railway bridges in the Victorian era of which an example is shown in Fig. 11.6. To quote from *The Illustrated London News* of 9 May 1891: 'The disaster on Friday May 1, at the Norwood Junction Station of the London and Brighton Railway, from the collapse of the iron bridge over Portland Road, when an express train was passing over it, might have had dreadful results. ... There was an undiscovered "latent flaw" in one of the girders of the bridge, which ought to have been reconstructed long since, as it gave way beneath a pilot engine fourteen years ago.' Such failures were a result of the poor tensile properties of cast iron in conjunction with defects which were more or less accepted features of casting at the time. Chapter 2 describes how from the experience of these failures arose a number of bridge designs employing cast iron in the compression members and wrought iron in the tension members. The latter material was not without its problems which were alluded to by I K Brunel in a letter to the commission appointed to enquire into the application of iron to railway structures. In his letter Brunel wrote: 'Who will venture to say that if the direction of improvement is left free, that means may not be found of ensuring sound castings of almost any form, and if twenty or thirty tones weight, and of a perfectly homogeneous mixture of the best metal?' Brunel's vision exists today in examples of cast steel nodes used in some offshore structures.

These accounts demonstrate that the leading engineers of the time were aware that metals needed more than tensile strength to support loads and that material flaws could affect the integrity of a structure. Even today most conventional structural engineering design procedures assume that the material and the joints contain no random imperfections which would prevent structures made of them failing to perform their task and that the mechanical properties are entirely uniform throughout the material. Earlier in this book we saw that welding processes, particularly in their manual forms, are subject to variations in behaviour which can result in unplanned variations and even discontinuities in a welded joint. Earlier chapters show that to cope with this situation there are weld defect acceptance standards, *workmanship standards*, which have grown out of common practice. These often represent what is achievable by good practice or defects which can be easily discovered but are in no way related to the effect of any weld flaws on the integrity of the structure. Their validity rests on the past satisfactory use of them in conjunction with controlled material properties which again have no theoretical relation to the tolerance of the allowable weld defects. In recent years, inspection techniques and operator training have improved so that it is possible to define the shape, orientation and size of an internal weld or material flaw far more accurately than in the past. Every now and then an engineer is faced with the problem of what to do with a fabrication defect which is larger than the specification would allow but whose removal would be difficult or expensive. Another matter is when a crack is found to have developed in service and it is necessary to decide if the crack will reduce the integrity of the structure and whether it is likely to grow. In both of these situations the engineer has to decide if the structure is fit for its purpose in the presence of the flaw or crack. The engineer can then turn to a procedure known as *engineering critical assessment* (ECA). When applied to the toleration of weld defects this involves making an assessment of the effect of the flaw on the integrity of the structure. This assessment is made by analysing the way in which the presence of the flaw modifies the local stress field and affects the potential for the propagation of cracks by brittle fracture, fatigue, stress corrosion cracking and so on. The approach makes use of the concept of fracture mechanics which was originally postulated by

G I Taylor and developed by A A Griffith[45] for explaining the behaviour of cracks in brittle materials. It has since been extended to be applicable to crack behaviour under non-linear stress/strain conditions such as exist in elastic/plastic materials including steels. The theory and the techniques are quite sophisticated and the satisfactory usage of the methods requires a fundamental understanding of the basis of the concepts and their inherent underlying assumptions. For this reason their use is best left to the fracture mechanics specialists. However in the early 1980s it was recognised that this was such a powerful and potentially beneficial tool that guidance on the use of fracture mechanics in assessing welded joints in respect of fatigue cracking and brittle fracture should be made available publicly. Such guidance was published in the UK as British Standard, PD 6493. A comparable document entitled *The Fitness for Purpose of Welded Structures* was published by the International Institute of Welding in 1990 as a draft for development but never published. An amended version of PD 6493 was issued in 1991 and a development of this was published in 1999 as BS 7910.[46] CEN, the European standards body, through its Technical Committee 121, is planning to issue the same document as one of its Technical Reports. Although they should not be used as textbooks, such documents as BS 7910 represent a condensation of knowledge of and experience in the application of defect assessment methods. They are designed to be used by people with some background knowledge of fracture mechanics and are generally conservative in their results. Nonetheless it is essential that the user ensures that the information used in deriving a decision on the acceptability of a certain defect is reliable. The two most common forms of fracture against which weld defects are assessed are fatigue and brittle fracture. BS 7910 sets out the assessment procedure in a number of steps.

For an assessment for brittle fracture, a knowledge of the fracture toughness of the material surrounding the defect is required. This can be as a critical stress intensity, K_c, or a critical CTOD type of measurement, δ_c. Several levels of assessment are offered in BS 7910 in increasing degrees of confidence accompanied by increasing computation and an increasing need for accurate materials and stress data. For assessment of a weld defect in respect of fatigue, two approaches are given. One equates the effect of the defect with the fatigue performance categories in BS 7608. The other requires the calculation of the history of the crack front growth by an iterative procedure. This can be quite a complex and time consuming exercise. As we have come to expect there is software[47] available which can perform these calculations.

12.1 What we mean by standards

The word 'standard' as it is commonly used in engineering is a contraction of *standard specification*. This is a specification for a material or manufactured product which may be written by companies for internal use, and by national and international bodies for public use. The word 'standard' also refers to *standard procedures* such as examinations and tests of materials and personnel. There are other types of 'standard' in a different context, for example the *standard metre* is the basic measure of length which was originally represented by the length of a platinum bar kept in Paris. Such basic standards have been replaced by more esoteric measures such as the distance travelled by light in a vacuum in a certain time.

12.2 Standard specifications

These have a number of purposes. At a simple level their use minimises the cost of production and maintenance of engineering products through the reduction in variety and the resulting interchangeability of similar parts. An historical example of the effect of lack of standardisation was in screw threads. Until the 1960s some countries used several thread sizes quoted in inches which included such forms as Whitworth, British Standard Fine (BSF), British Association (BA), Unified Coarse (UNC) and Unified Fine (UNF). Some manufacturers even had their own threads such as were used in the BSA (Birmingham Small Arms) bicycle. The owner of one of these bicycles had to make sure that any replacement nuts or bolts were to the BSA thread. Eventually metric sizes were adopted by most countries and matters became much easier to manage both in factories and in the customers' maintenance departments. The reduction in variety offered lower costs through increased production runs of parts, reduced stock holdings of finished parts and fewer types of tools used in manufacture and maintenance, for example taps, dies and spanners. Standard parts also save design time whether they be fastenings, couplings or cable terminations, because all that needs to be done is to call up the standard number as the part number on the assembly drawing.

Products made to a standard specification can be used with others in different countries, regions or continents. We can buy the same torch batteries, photographic films and car fuel over most of the world thanks to standards but strangely enough we still can't always use the plug on our electrical equipment all over the world. In the business of information technology we still have the ridiculous situation where we have mutually incompatible software. The text of this book cannot be simply transferred to some other PC which does not have the same word processing software unless there is a conversion program installed. We cannot even copy text to a floppy disk and automatically expect to be able to run it on a different make of PC even with the same software. The 'floppy' disk itself is identified by its diameter measured in inches and not millimetres which is the international unit because in the USA, where the floppy disk was first marketed, the international system has not been implemented, about the only country in the world not to do so. Perhaps in a few years standardisation will reach the information technology industry; the potential savings in time and cost will even now be evident. As recently as 1999 a spacecraft failed to land intact on the planet Mars, evidently because instructions had been given in miles and not kilometres, a truly costly example of the consequences of a failure to use standards.

A second purpose of a standard is to describe a product which has a specific level of performance. This is of particular importance when, for example, a standard specification includes provisions for safety. Standards can ensure consistent performance of engineering products which require to be designed to a particular philosophy on a basis of reliable performance data. Such design methods and data often come from a variety of sources and over a long period of time. The function of a standard specification is often then to provide a digest of this data which will have been assessed for validity to ensure that it can be used reliably within the context of the standard. Examples are seen in product standards such as those for bridges, chemical process plant, offshore platforms and cranes. Such standards are more difficult to compile than a description of a part such as a bolt and cannot specify the eventual product as a physical item. They have to be more correctly thought of as codes of practice, leaving the engineer free to design and manufacture the product as he thinks fit whilst conforming to the intent of the standard.

A standard must be written not only so as to define the characteristics of the product but to define how that product will be demonstrated to conform to the standard. This is feasible where, for example, material composition and strength or the dimensions of a screw thread are specified. However when we get to something as large as a building or a bridge how do we demonstrate conformance with the standard? One way of course is to perform a completely separate set of calculations from the original design calculations. Another is to put loads on it and measure the stresses or to measure stresses in service.

It must be said that standards should be used only as a support for good engineering and not as its basis. Standards are derived jointly by the parties interested in making and using the product as well as by others and the time taken to prepare and publish a standard means that it cannot be based on the most up to date technology. The result can then represent only rarely anything except the lowest specification acceptable to those parties. In any particular application, the standard alone may not represent all the requirements of the customer or the manufacturer and the creation of a sound engineering product requires that it be specified, designed and manufactured by competent engineers. There is a view that the availability of, and adherence to, detailed standards is not necessarily beneficial because, as we have claimed above, engineers can use them as design aids instead of seeking new solutions to requirements thereby discouraging the development or adoption of more advanced approaches. Furthermore the existence of detailed standards makes it possible for people of little familiarity with the subject to attempt to design and manufacture products about which they know little. It must be emphasised that the application of a standard requires that the user understand the circumstances for which it was prepared; it can be very dangerous to use standards in ignorance of their derivation and scope.

In the field of welded fabrication there are many standards describing the materials, welding consumables, welding plant, the management of welding operations, inspection techniques and procedures and the fabricated product itself. Naturally many of these standards will be called up by manufacturers and customers in their product specifications, unfortunately not always with adequate knowledge of their scope and content. There are in existence many in-house company specifications which have been used for years during which time they may have been amended by people without specialist welding knowledge to suit new jobs and for which the originally quoted standards are inappropriate or which even conflict with the basis of the design. This is a circumstance where the welding engineer will be needed to advise on the interpretation or even the rewriting of the specification.

It is an unfortunate fact of life that the first act of many writers of project specifications is to reach for the list of standards. This should of course be the last thing that they do. The first and most important matter to be decided is the basis of the design. That is to say what the product is intended to do, in what way will it do that and what means of realisation will satisfy that. In practice, and depending on the particular industry, the specification will be more or less detailed, and will call up such standards as are technically appropriate or as are required by legislation, the customer or other authorities. Project specifications will go through several stages as will the design; for example in civil, structural and other heavy engineering these stages may include a feasibility study, conceptual design, design specification, detail design, fabrication specification, fabrication or shop drawings, design report and finally the as-built records.

There are standards which are applicable to all the subjects of the chapters in this book from product standards such as bridges, buildings, cranes and pressure vessels, welding materials such as welding equipment and electrodes and techniques such as non destructive examination. On an international scale standards are published by ISO and IEC; on a regional scale there are standards such as those published in Europe by CEN and CENELEC. On a national scale there are standards published by national standards bodies as well as by professional institutions, commercial bodies and individual manufacturers. In general access to international and regional standards in any country is through that country's national standards body.

References

1 Newman R P, 'Training for welding design', *Improving Welded Product Design Conference*, Abington, The Welding Institute, 1971.

2 Cottrell A, *An Introduction to Metallurgy*, 2nd edn, London, Edward Arnold, 1975.

3 *Training in aluminium application technologies, TALAT CD-ROM Version 2.0*, Brussels, European Aluminium Association, 1999.

4 *Glossary for welding, brazing and thermal cutting*, BS 499: Part 1: 1991, London, British Standards Institution.

5 *Welded, brazed and soldered joints – symbolic representation on drawings*, ISO 2553, Geneva, International Organisation for Standardisation, 1992. (Also as EN 22553 published under their own prefixes by the national standards bodies of CEN.)

6 *Welding, brazing and soldering processes – vocabulary*, ISO 857, Geneva, International Organisation for Standardisation, 1990.

7 *Standard welding terms and symbols*, A 3.0, Miami, American Welding Society.

8 Tsai K C, Chen C-Y, 'Ductile steel beam-column moment connections', Proc, *IIW Asian Pacific Welding Conference*, Auckland, New Zealand, 1996.

9 'Design rules for arc-welded connections in steel submitted to static loads', Doc XV-358-74, *IIW*, 1974 (unpublished).

10 Clark P J, 'Basis of design for fillet-welded joints under static loading', *Improving Welded Product Design Conference*, Abington, The Welding Institute, 1971.

11 *Steel structures, Part 1: Materials and design*, ISO 10721-1, Geneva, International Organisation for Standardisation, 1997.

12 Wohler A, 'Tests to determine the forces acting on railway carriage axles and the capacity of resistance of the axles', *Engineering*, 1871, **11**.

13 Shute N, *No Highway*, London, Heinemann, 1949.

14 Gurney T R, *Fatigue of Welded Structures*, 2nd edn, Cambridge, Cambridge University Press, 1979.

15 Gurney T R, *The Basis of the Revised Fatigue Clause for BS 153*, London, The Institution of Civil Engineers, 1963 (*Discussion*, 1964).

16 Signes E G, Baker R G, Harrison J D, Burdekin F M, 'Factors affecting the fatigue strength of welded high strength steels', *Br Weld J*, 1967 **14** 3.

17 Maddox S J, *Fatigue Strength of Welded Structures*, 2nd edn, Abington, Abington Publishing, 1991.

18 Gurney T R and Maddox S J, 'A re-analysis of fatigue data for welded joints in steel', *Weld Res Int*, 1973, **3** (4).

19 Pilkey W B, *Peterson's Stress Concentration Factors*, 2nd edn, Chichester, Wiley, 1997.

20 Marshall P W, *Design of Welded Tubular Connections*, Amsterdam, Elsevier, 1992.

21 *Structural Use of Aluminium*, BS 8118, London, British Standards Institution, 1991.

22 Honeycombe R W K, *Steels, Microstructure and Properties*, London, Arnold, 1981.

23 Boyd G M, *Brittle Fracture in Steel Structures*, London, Butterworths, 1970.

24 Tipper C F, *The Brittle Fracture Story*, Cambridge, Cambridge University Press, 1962.

25 Rolt L T C, *Victorian Engineering*, London, Pelican Books, 1974.

26 Dwight J B, 'Effect of welding on compression elements', *Improving Welded Product Design Conference*, Abington, The Welding Institute, 1971.

27 Young W C, *Roark's Formulas for Stress and Strain*, Basingstoke, McGraw-Hill, 1989.

28 Wardenier J, *Hollow Section Joints*, Delft, Delft University Press, 1982.

29 *SHS Welding*, TD 394, 15E.97, London, British Steel, 1997.

30 *Front Line*, London, His Majesty's Stationery Office, 1942.

31 Baker J F, *The Steel Skeleton, Vol 1, Elastic Behaviour and Design*, Cambridge, Cambridge University Press, 1954.

32 Baker J F, Horne M R and Heyman J, *The Steel Skeleton, Vol 2, Plastic Behaviour and Design*, Cambridge, Cambridge University Press, 1956.

33 *Steel Designers' Manual*, London, Blackwell Science, 1994.

34 Davies J M and Brown B A, *Plastic Design to BS 5950*, Oxford, Blackwell Science, 1996.

35 *Report of the Inquiry into the Causes of the Accident to the Drilling Rig 'Sea Gem'*, Cmnd 3409, London, HMSO, 1967.

36 'Giant Offshore Structures – Whose responsibility?' *Offshore Services*, Vol 6, No 5, Kingston-upon-Thames, Spearhead Publications, July 1973.

37 Hicks J G, *A study of material and structural problems in offshore installations*, Welding Institute Research Report E/55/74, January 1974.

38 Hicks J G, 'A study of material and structural problems in offshore installations', *Welding and Metal Fabrication*, 1974, **6** (9).

39 *Offshore Installations: Guidance on design, construction and certification*, London, HMSO, 1990 (Revoked 1998).

40 Reid A, *Project Management: Getting it Right*, Cambridge, Woodhead Publishing, 1999.

41 Burgess N T, ed, *Quality Assurance of Welded Construction*, 2nd edn, London, Elsevier Applied Science, 1989.

42 Harrison J D, Burdekin F M and Young J G, 'A proposed acceptance standard for weld defects based on suitability for service', Proc 2nd conf, *Significance of Defects in Welds*, London, The Welding Institute, 1968.

43 Lancaster J, *Metallurgy of Welding*, 6th edn, Abington, Abington Publishing, 1999.

44 Lancaster J, *Handbook of Structural Welding*, Abington, Abington Publishing, 1992.

45 Griffith A A, *Phil Trans*, A-221, 163–8, London, Royal Society, 1920.

46 *Guide on Methods for Assessing the Acceptability of Flaws in Fusion Welded*

Structures, BS 7910: 1999, London, British Standards Institution, 1999.

47 *Crackwise 3, Automation of BS 7910: 1999, Fatigue and fracture assessment procedures* (software on disk), Abington, TWI, 1999.

Bibliography

Some publications which may be a useful background or provide further references:

BS 7608, *Code of practice for fatigue design and assessment of steel structures*, London, British Standards Institution, 1993.

Recommended practice for planning, designing and constructing fixed offshore platforms, API RP 2A, Washington, American Petroleum Institute.

Specification for the use of structural steel in building, BS 449: Part 2, London, British Standards Institution, 1995.

Structural use of steelwork in building, BS 5950, London, British Standards Institution.

Steel girder bridges, BS 153, London, British Standards Institution, 1958.

Steel girder bridges, Amendment no 4 to BS 153: Part 3B and 4, London, British Standards Institution, 1962.

Steel, concrete and composite bridges, BS 5400: Part 10: 1980, *Code of practice for fatigue*, London, British Standards Institution.